BORN 150 YEARS TOO LATE

The Musings of a Modern Wilderness Junkie

BORN 150 YEARS TOO LATE

The Musings of a Modern Wilderness Junkie

JEFFREY YORK

SUNSTONE PRESS

SANTA FE

Sunstone books may be purchased for educational, business, or sales promotional use.
For information please write: Special Markets Department, Sunstone Press,
P.O. Box 2321, Santa Fe, New Mexico 87504-2321.
Printed on acid-free paper
∞
eBook 978-1-61139-717-8

Library of Congress Cataloging-in-Publication Data

Names: York, Jeffrey, 1958- author.
Title: Born 150 years too late : the musings of a modern wilderness junkie
   / Jeffrey York.
Description: Santa Fe : Sunstone Press, [2023] | Includes bibliographical
   references. | Summary: "A nonfiction book about outdoor adventure along
   with environmental discussions on the American West"-- Provided by
   publisher.
Identifiers: LCCN 2023032994 | ISBN 9781632935441 (paperback) | ISBN
   9781611397178 (epub)
Subjects: LCSH: Wilderness areas--West (U.S.) | Outdoor life--West (U.S.) |
   Diet--Environmental aspects. | Ranching--West (U.S.) | Hunting--West
   (U.S.) | LCGFT: Creative nonfiction.
Classification: LCC QH76.5.W34 Y67 2023 | DDC 333.720978--dc23/eng/20230724
LC record available at https://lccn.loc.gov/2023032994

WWW.SUNSTONEPRESS.COM
SUNSTONE PRESS / POST OFFICE BOX 2321 / SANTA FE, NM 87504-2321 /USA
(505) 988-4418

In memory of my father Norman Leon York who led his little son into the ancient classroom, and showed him the beauty that dwells in all things.

CONTENTS

INTRODUCTION

A warning: this book is not a hiking guide. It is not intended in any way to inform the reader about hiking areas, trails, or routes. This book gives no information on how to get there, what to wear, and what to expect. There are no maps herein and no cute little lists of dangerous things to watch out for.

You don't need one of those books anyway. Throw the accursed hiking guide into the flames and just go! Climb a cliff. Swim a wild river. Find an arrowhead. Smell the sundown wind. Drink from a hidden spring. Spook a moose. Startle a bear. Sprain your ankle. Get bitten by a large, carnivorous insect. Get totally lost. Feel the lovely pang of despair. Bleed. Sleep in the dirt under a perfect sky. Experience complete and total terror. Dig for water. Sit in the smoke to escape the mosquitoes. Get wet, cold, and miserable. Scale a big mountain, just for the view. Watch a bighorn ram sprint past you. Have an adventure. And whatever you do, when you find one of those precious few remaining gems of wildness, protect it and cherish it. Hold it firm and high in your memory. And don't tell anyone where it is.

I started exploring Arizona and the American West in the 1960s, and have poked my nose into a lot of untamed nooks and crannies. What started out as a mild childhood interest in things beyond the city limits grew into a lifelong passion, and walking across wild country increased my appetite to see more.

But something else was happened as well. While exploring, I was witness to a terrible and unrelenting destruction of the American West. Although the past three decades have brought many victories for environmentalists, the overwhelming number of people pouring into Arizona and every other state in the West makes every green triumph seem like tossing a stone into a torrent. Arizona and her sister states in the West are being gobbled up at a rate that can only be called frightening.

Fascinated with wilderness, I have in some dark fashion become mesmerized with its destruction. Under the unrelenting onslaught that gains momentum as the years pass, it is astounding that any area has managed to say wild and clean.

When I was young, my father told me, "Son, you were born a hundred and fifty years too late." For the most part, I agree, because I missed the frontier. At the same time, I feel fortunate not to have been born a hundred and fifty years from now. At the present rate of human encroachment there won't be enough wild land to walk across in half a day.

So then, what is an outdoorsman's story if it's not a how-to, if it's not a two-hundred- page listicle, or an environmentalist manifesto? This book is about love, I guess. Because if we love the wild land hard enough, it won't disappear.

Throw Your Hiking Guide into the Flames! (Illustration by Jan York)

Thunderstorms and Sandstone: A Prologue

After several days alone you start talking to trees.

I haven't spoken to a living, breathing creature in five days, unless you count the rattlesnake. I had thanked him (Her? And how would you know?) profusely for not biting me when he had every opportunity and every right to bite me. After all, I put my left foot down on the snake and then stood on him for a while. Looking down, I saw the snake squirming to get out from under my foot and noticed the buttons on his tail and the wicked wedge-shaped head. A fraction of a second later, I was two feet in the air and five feet to the side, backpack and all. The snake crawled into the brush and coiled up, looking at me with its beady eyes. Lying sprawled on the grassy slope, I looked back with my own beady eyes and thanked him. Taking a snakebite eight miles from my truck in the middle of the Arizona summer is not my idea of a pleasant experience.

Rattlesnakes have my respect. They are fierce, well-armed little warriors. I don't walk around in fear of them, but many years of experience and a couple of close calls have trained my eyes to keep a constant, if mostly subconscious, vigil for snakes. How I walked up and stood on one is beyond me. The problem with snakes is that they are difficult to see if they are not moving. Their natural camouflage blends with a broken background of twigs, grass and leaves very well. A mistake like that can cost you dearly. Nice snake.

The sun is intense at this altitude, but it is a cool eighty degrees in the shade. Five thousand feet below, the Sonoran Desert shimmers and shakes, baking at a hundred and ten degrees. To the south, a massive wall of storm cloud is sweeping up from Mexico. The Huachuca Mountains, a tall sky island massif very similar to the one I am sitting on, have been blotted from view by rain and cloud. If the big cell holds together it will reach me in another hour or two.

Why do I do this? My friends are at parties or the mall. Watching football games on T.V. Washing the car. Doing things that normal people do. And here I am, sitting on the edge of a cliff near the top of a huge sunbaked ridge, waiting for a thunderstorm. Eating freeze-dried food, sleeping in the dirt, and talking to snakes. Breaking all the major rules of Safe Hiking Conduct:

1. Never hike alone.
2. Tell people where you are going, your plans, and when you will return.
3. Stay on the trail.

The storm is definitely going to reach these mountains. A thin veil of cloud, torn from the towering anvils of the thunderheads, is streaming overhead, slightly softening the sun's power.

Carefully backing away from the edge of the cliff, I turn and walk down a brushy, rocky slope into the small basin behind the ridge. Tucked into a tiny stand of Ponderosa pine in the center of the basin is my pup tent. Reaching inside, I pull out my slicker. While heading back up the slope with the slicker draped over my arm, my mess kit catches my eye. I had better straighten up this camp. If I am lucky and the storm does come through this area, my stuff will be blown around and thoroughly soaked. I gather up everything in camp, including my backpack, and put it inside the tent.

"Inside" is a relative term when describing the interior of a pup tent. A pup, or "tube" tent is nothing more than a cylinder of thin plastic suspended from a line tied to a couple of trees. When sleeping in a pup tent in moderate to serious weather, it is often difficult to determine where the "outside" of the tent ends and the "inside" of the tent begins.

Returning to the cliff, I sit on the edge facing south. Across the valley to the southeast, the smooth domes of the Mustang Mountains have disappeared behind a racing black wall. To the east, rain falls in long windblown streamers of water that swirl like gossamer scarves. The rain evaporates in the hot dry air before reaching the ground.

The basic problem here is that I'm addicted. I've got a serious problem that they don't handle at clinics. I'm a wilderness junkie. It's not a mild habituation, like a craving for a cup of coffee in the morning. This is a deep, wild need. You won't find needle tracks on my arms, but you will find the telltale signs of my habit, the brush scratches and the insect bites, the sunburned peeling skin, the lost faraway stare. My house is littered with paraphernalia; backpacks, canteens and the maps. The maps, my beautiful maps!

I'm so far gone I don't even try to hide it.

That is why I am here. I have to be. There is really no choice. I'm not one of those people who say that they just love nature, and then go hiking a full two times per year for a total of about four hours and then only when the weather is perfect. I have to get my regular fix. If not, I will go into withdrawal. First comes the writhing, then the moaning, followed by the pathetic stage of staring at old outdoor photos and drooling on maps, and then finally dropping hard into the bleak condition when the sight of one more span of asphalt, one more hideous strip mall, one more freeway on-ramp, one more polluted sky, one more demon-spawned hell bent motor home will drive me into an extended foaming spasm.

It started when I was a boy. You could call it the family curse, as it seems to have been passed down from generation to generation. Dad would notice my hungry stare as we sat together, high on a lonesome ridge, just taking it all in. He would tell me, laughing sometimes and sometimes not, that we were both born too late. We had missed the frontier. His father had missed it, too, but his grandfather, that wild old man, had "seen the elephant," the untamed frontier.

Even so, the curse was controllable until my father took me to southern Utah for a fishing trip. He and I and my uncle camped on the shore of Lake Powell for a couple of weeks. Fishin' (no one who goes fishin' actually says "fishing") seemed to be a very manly activity, involving gear, weather, terminology, the slaying of critters, and camping out. Being ten years old and very interested in the mysterious arts of manhood, I plunged into fishin' with the gusto that only a ten-year-old boy can bring to the sport. And did we ever catch fish. Fishing on Lake Powell was in its heyday at that time, when the lake was young and the striped bass hadn't yet been introduced into the lake. In a single day you could catch largemouth bass, smallmouth bass, crappie, trout, maybe a catfish or two, and even northern pike.

Those were magical days, boating the big lake with the two veteran outdoorsmen. We would cross a big bay, the small boat bouncing over the waves whipped up by the incessant wind, and hunt along the shore for a narrow side canyon. When we found one, we would putter along with the outboard engine at idle, looking for submerged trees. A grove of drowned trees was almost certain to be a prime fishing spot. I would drop a bright yellow jig into the clear water and watch in delight as the crappies rose from the depths to attack the lure. In one afternoon, we caught over two hundred crappies. Dad and my uncle filed the barbs from the hooks of my jigs, and sat back and laughed as I pulled fish from the lake as fast as I could cast and reel them in. Many times, I would catch and release a fish, and then immediately catch the same fish again. During the peak of this fish madness I tied a bare treble hook onto my line, weighted with a couple of split shot halfway up the leader. On the second cast I pulled three crazed crappies out of the water.

Sometime during that long sunny October afternoon, I realized I had never been happier in my short life.

At the end of that day we beached the boat on the submerged peak of a once mighty sandstone pillar, now reduced to a rocky island several hundred yards from shore. It took us a full hour to clean and filet our catch. That night we feasted on fresh crappie and bass broiled over a priceless cache of mesquite wood that had been transported from a grove near the Mexican border.

In the mornings with the coming sunrise still a blush in the sky, we would be a mile out in the lake, huddled down to escape the cold breeze coming over the bow as we cracked across the water at full throttle. By the time the sun came up at least one of us would have landed a fish.

My hands tore and bled, and then toughened and hardened under the caress of line and barb and the sharp fins of the big bass. I learned to tie a blood knot and filet a crappie. My father taught me how to work a surface jig and a bass worm. My small quick fingers became so adept at tying the intricate knots and working the tiny swivels that the two men soon had me rigging their lines. My uncle showed me how to take a small open boat across a big bay against a howling sandstorm, with the sand hissing into the water like rain.

The enchanted days went on and on; full of golden light and clear water, of red rock and laughter, and at the age of ten years I became a fisherman. I could talk fishin' with the men, hunkered down on a weathered dock with my little hands pantomiming the motion of the lure, the streaking attack of the fish, the strike, and the glorious fight of the bass. The big bearded men would nod solemnly when I sadly mentioned the whale of a largemouth bass that broke my eight-pound test line right underneath the boat. This was at the very height of the Cold War and we were preparing to put a man on the moon, but that was small stuff. This was serious business. This was fishin'.

I thought I would burst with boyish pride one afternoon when a man who was obviously impressed with our bulging creels asked my father what lures he would recommend for crappie. Without hesitation, my father referred the man to me.

"My son here is the crappie expert. He can catch crappie when no one else can," he said, motioning to me with his pipe.

Sitting in his very expensive bass boat loaded with about three thousand dollars' worth of fishing gear, the man asked me about my crappie fishing techniques without a hint of condescension.

Trying to maintain a matter-of-fact attitude, I replied, "I would go with a small yellow jig, with a couple of split shot tied about three-four inches up the leader. The extra weight gives me a bit more accuracy when casting. I like to work the jig as close as I can get to a tree or bush, you know, submerged trees. I work 'em like this, to make the tail dance."

The man watched my short cast and the bright jig jumping through the water a few inches under the surface.

"Yeah, that's a nice action. Just reel it in real slow, eh?" he asked, grinning past me at my beaming Dad.

"Yeah, like that. And remember to work close to the trees. You'll lose a few jigs that way, but you'll catch a lot of fish!"

With a grave "Thank you, sir," and another smile at my father, he moved off down the lake.

I had pretty much reached the pinnacle of existence at that point, or so I thought. I told my uncle and my father over dinner one night that I was going to be a professional bass fisherman when I grew up. I mean, what could be better?

And then one afternoon I went for a walk. It was a slow, lazy day on the beach. The boat motor was acting up and my uncle had it torn into about twenty pieces. My father had his head underneath the hood of his truck, and neither man was talking. They were both deep into the world of tools and grease and machinery. I thought about offering my help, but wisely thought better of it. Just behind our camp was a sizeable pink sand dune, fifty or sixty feet tall. I climbed it, and right there on the other side was paradise. The rock looked like candy, red and pink and striped and stained. There were dozens of sandstone spires standing like giants in a fantastically shaped rock basin, themselves dwarfed beneath enormous smooth domes of sandstone hundreds of feet high.

I looked back at camp, down at the two men engrossed in their tasks.

"Goin' for a walk, okay!" I yelled.

Without looking up, my father waved me on.

I was free.

Plunging off the dune, I charged out onto the slickrock. It just didn't seem possible to me that all this rock could be standing so vertically. It looked like it should all fall down. I darted among the goblin spires, running my hands over the smooth sandstone. Scrambling from ledge to ledge, I climbed onto a massive dome of red rock. The rock was nearly seamless and very smooth for hundreds of feet, and up I went. The top of the dome was pitted and pocked, with small pools of water lying in the natural basins. From below, it looked as though the dome was a peak, a high point, but when I got to the top I was looking across at another dune, which was itself at the base of a long slope of sand and broken stone. The slope went up another several hundred feet, ending at a hard cap of white sandstone, the edge of a small mesa. Beyond and above the mesa, the rock bulged and twisted into a sheer wall, itself the edge of yet another mesa. Above this was another steep broken slope, and then still more ragged ramparts, and beyond them, now thousands of feet over my head, were the five hundred-foot cliffs that marked the eastern edge of the Kaiparowits Plateau.

I could not believe my eyes. From a distance, it had looked as though the land rose to the plateau in one long rough leap. Only a closer inspection revealed its incredible complexity. I sat down and faced northeast, back toward the lake. From my high vantage I could see the drowned canyon under its sparkling blue shroud. Beyond the lake, buttes and towers and mesas for untold miles, and the great blue monolith of Navaho Mountain ruling the twisted crumpled raw skeletal land. I loved it.

Even then, I might have escaped the curse if I hadn't found the slot canyon.

The mouth of the canyon was so well hidden that I would not have found it if I hadn't been right in its watercourse. The dry wash was an excellent place to find pretty rocks, and I was meandering along, picking up and admiring one rock and another. The wash puzzled me. It looked as though it ended at the base of a sandstone cliff just ahead, but I knew that couldn't be. There must be a waterfall, I told myself. Walking along the base of the cliff looking for the worn channel that water would come pouring down when it rained, I felt a cool breeze flowing along the face of the cliff. Strange. The cool air was flowing out of a crack in the sandstone. Rounding the end of a narrow fin of rock that swooped down from the cliff like a big wing, I found that the crack was the mouth of a canyon. The canyon was about twenty feet wide where it exited the sandstone, and perhaps fifty feet deep. The canyon bottom was of smooth sand. It was very inviting, and I did not hesitate.

Once inside the canyon I was in another world. The canyon walls were fantastically shaped, bulging out from the rim in improbable defiance of gravity, fluted and pocked here, and there as smooth as glass. Stains rich and varied hung down the red stone like fossilized tapestries. As I walked upstream the canyon cut deeper and deeper into the rock, a sinuous snakelike groove weaving its way into the ancient stone. The bright sunshine rarely reached the bottom, but occasionally broke through to pool on the walls above as I crept along under the overhanging rim. Each bend of the canyon revealed some new miracle or another rare view. One side of the canyon would bulge out, and the other side would be concave, each matching the other perfectly. At one point, a deep break in the rimrock allowed the sun to shine directly onto the canyon floor, and in the center of the light was a perfect little tree. Its bright green leaves were in soaring contrast to the deep reds and pastel pinks of the stone. As I approached the tree, I got just the barest hint of the sweet scent of water. Placing my hand on the dry sand at the foot of the tree, I could feel the cool presence of hidden water.

After I had gone a few hundred yards into the slot, the sky was suddenly not visible. The towering walls leaned overhead, blocking out the world above. The sunlight that reached the bottom had to bounce from wall to wall, giving the bottom of the canyon a strange orange glow. I cast no shadow. Scrambling over boulders that choked and plugged the canyon in places, I went up and up until my way was blocked by a huge boulder that had spalled from the canyon rim and wedged itself tightly into the bottom of the gorge. At this point the narrow strip of sand I stood on was only three feet wide. Looking up, I could see a tiny ribbon of blue hundreds of feet over my head. I was in a cave with skylights.

I played among the hoodoo rocks until it was dark. Standing on the crest of the dune, I waved goodbye to my friends, the goblins and dragons standing sentinel before the mouth of the secret canyon.

Goblin Spires. (Illustration by Jan York)

After that, fishing just didn't have the allure it had before. Sitting out in the boat, my eyes would turn to a ragged wall of cliffs, a noble mesa, or the mouth of a canyon hanging five hundred feet above the lake. I would wonder what was behind that mesa, and I found that I could fantasize for hours about the mysteries of that canyon up there that I could never hope to reach.

The men would shrug and laugh, and then my uncle would fire up the motor. After dropping me off on the shore they would turn about for more fishing. I would stand and watch as the boat dwindled to a dark spot on the water, then turn and spend the day walking between the standing rocks. It was during these long hours alone that I discovered that I love to walk across the earth. It is so simple and satisfying to propel yourself, not with an engine or some other contraption, but with your own body, across the ancient rippling contours of ridge and valley. Another important discovery was that I like rock and sand under my feet, not concrete.

One of the many extraordinary things about those last few days at the lake is that there were many places I explored where I may have been the first human to walk the land. The idea of this fascinated me. The shorelines I followed were very remote. I saw no footprints. No garbage. There was no sign that anyone had ever lived on these desolate benches of sandstone. Only a few years before the massive dam had not been down there across the southern end of Glen Canyon, and the river ran free. Many of the canyons and valleys I walked in had been hanging high above the river, protected from below by hundreds of feet of vertical rock. With the lake filled nearly to maximum it was easy to pretend that the boat was a helicopter hovering above the Colorado River, taking me to the wild places where no one had stood before.

§

The huge storm moves off the Huachucas and hits the base of the big ridge with a howling roar. The desert is obscured by dust and rain. Two thousand feet below me, I can see trees and brush flattened by a towering surge of air, moving fast and carrying sand and dirt and small debris. Suddenly I can hear it coming, hissing and spitting as it rolls up the slope. The cool mountain air around me is suddenly torn away by a hot blast of wind. The storm pushes hot air from the desert floor right up the mountainside at about seventy miles an hour. I jump up and run down into the basin, looking for shelter. The spindly Ponderosa pines are suddenly whipping wildly and the hot dry air is full of sharp flying things as this invisible tidal wave screams over the cliff. I throw myself down behind a large fallen log and cover my eyes with my slicker.

KA-WHACK!

Like a cannon shot, lightning hits the ridgeline a few hundred yards to the south.

My poor pup tent isn't long for this world. It is being torn to pieces.

As suddenly as it came, the hot dry wave is gone, and the wind is cold and damp.

WHAM! WHAM! WHAM!

Lightning works up the ridgeline toward me. The deafening thunder slams off the barren rocky slopes of the higher main spine of the range, rising just a few hundred yards to the west. Pulling on my slicker, I am undecided as to what to do next. I've got to move now, or ride the storm out here in this relatively protected basin. The thought of crossing over the barren cliff top is not very appealing; every big tree around the basin is scarred by lightning. On the other hand, I've been up here for five days waiting for a storm.

Jumping up, I trot up the slope. There, just to the right of the little barrel cactus. I slide down into the narrow break in the cliff, trying not to tear my slicker on the sharp rocks, and scramble down ten feet to a rock ledge

I had discovered a couple of days before. Made it! And here it comes, my incredible storm, right at me.

The ledge is a perfect viewing platform. It is tucked under the lip of the big cliff and the view to the south is unrestricted. At the base of the cliff, the ridge drops steeply down to the desert, and it feels like I am suspended thousands of feet in the air. But the ledge offers no shelter. It is simply a flat slab of rock with no overhang or protected lees to hide behind, but then again that is the whole idea. I will sit below the cliff top, safe at least from lightning strikes, and take the storm right in the face.

It won't be long now. The south end of the range is gone, hidden behind a dense curtain of rain and hail. Twenty miles to the east the Whetstone Mountains are shining in the sun, but they too will soon be gobbled up. This monster is big. The foot of the storm has to be forty miles across, maybe fifty; a twenty-thousand-foot cliff of falling water laced with bolts of lightning and rolling north at about forty miles an hour. Thirty or forty thousand feet overhead, the crown of the storm is a roiling angry mass of raw energy. The clouds flicker as lightning leaps from crown to crown.

Thunderstorm. (Illustration by Jan York)

The wind dies down, gusting now and again. Even though it is mid-afternoon, it is as dark as dusk. Big raindrops begin to spatter on the warm rock. There is so much lightening that the sound of thunder becomes a continuous roll of mighty drums.

The sweeping view is suddenly gone as the fantastic amount of energy stored in the storm cell begins to lash the side of the mountain. Waves of hail and rain pour down out of the raging sky. Lightening pounds the ridgeline, bolts hitting so close that the flash and roar are simultaneous.

The wind is suddenly incredible. It must be blowing fifty or sixty miles an hour. Frustrated by the cliff, it cuffs me from every angle like a big wet angry bear. I huddle into my cheap slicker and dig both hands into cracks in the rock to keep from being blown off the narrow slippery ledge. The wind finally finds freedom a few feet above my head, shrieking over the shattered cliff top like a fleeing demon. The rain comes down so hard that visibility is reduced to a few yards, and with the rain hitting me hard in the face I can't see more than a few feet. I get a weird feeling of vertigo when the chasm in front of me is replaced by this moving waterfall, and right at the peak of all this chaos, an enormous bolt of lightning shoots down past me into the canyon thousands of feet below. Half crazed with fear and ecstasy, I am howling like a wolf into the teeth of the storm.

§

The fishing trip to Lake Powell finally ended, and it was time to leave the goblins and dragons to their timeless vigil and return to the world. To enter Arizona, we had to drive across the bridge that spans Glen Canyon. The bridge runs directly over Glen Canyon Dam. On the drive up to the lake I had been enthralled with the size of the dam, and proud that my father had helped build it. And now I hated it. It had buried the beautiful wild canyon. My father had been one of the last people to float the Colorado through Glen Canyon before the river was killed, and I admired and envied him for that lost, unrepeatable experience.

When we left Lake Powell we drove south along the grim, ragged ramparts of the Echo Cliffs, near the northern border of Arizona (Nearly four hundred miles to the south, the Huachuca Mountains, which neighbor the peak where I waited for the thunderstorm fifteen years later, rise at the very southern reach of Arizona, at the border with Mexico).

To the west, the great red wall of the Vermillion Cliffs rises two thousand feet above its ancient feet. Between the two lines of enormous sandstone battlements runs Marble Canyon, a cleft so deep and narrow that from a distance of just a few miles it looks as though it was cut with a skyscraper-sized knife. The dazzling light, the vast sweeps of desert, and incredible size of everything was stunning.

"That boy is gonna twist his head right off trying to see everything, Norman," my uncle chuckled. "Nothin' out here but sand and rock and wind, Jeff."

My father laughed, and said something about a godforsaken hellhole.

An abandoned Navajo hogan sat forlorn and lonely at the very foot of the desolate cliffs.

"I never want to leave here," I murmured to myself. "I want to live here forever."

Seventy miles to the south the highway leaves the desert, rising gently but steadily up the long, long slope that leads to Flagstaff and the volcanic wilderness of the San Francisco Peaks, thirteen thousand feet high. Part way up the slope, I made them stop the truck. I walked back along the highway and just stood, looking back and down at the tumbled desolation, the beautiful wasteland. I tried to make myself like a sponge so I could absorb it all, all the light and space and color and heat and vast emptiness. The two men waited, quietly and very patiently, while I said goodbye to the desert. When we finally drove on and the desert was lost to view behind the pine forest, I felt a longing that has never left me, and never will until the moment I die.

I can close my eyes now and see the light over Gray Mountain.

DESERT DREAMS

Here it comes, on demon wind,
Behold it now, the devil's friend.
From crushing foot
To anvil crown
It comes to tear the mountains down.

It is early September and the peak of the summer rain season. The thermometer in the shade on the back porch reads 108 degrees and the air over Tucson just drips. The swamp cooler on the roof chugs away, managing to lower the temperature of the air across the pads by approximately five degrees.

I lean on the accursed push mower (economy model, $39.95) and stare at the heavy, dark clouds hanging over the Santa Catalina Mountains. A hurricane or tropical storm, some sort of huge natural engine whirling madly away over the Pacific, has pushed immeasurable tons of airborne water into the southwestern U.S. The combination of moisture and Arizonan heat creates magnificent and often violent local thunderstorms. It is my favorite time of the year, what we locals call the "Monsoon Season."

From the front yard I can see a thick ribbon of what appears to be a solid stream of water pouring into a canyon on the Front Range. Closing my eyes, I imagine myself up there in Pontatoc Canyon, with the cold rain, born twenty thousand feet above my head, glancing from steaming stone, cactus, tree, lizard and coyote, drenching my shirt as I stand and face the storm. Desert dreams.

There is no relief here. The sun shines down unobstructed, free to bake my brains out. Sweat pours from underneath my sodden hat, runs down my face and chest, rolls over my ribs, and trickles down the small of my back. Pushing the mower through the thick grass, I decide that the hedge-trimming job can wait. The arrival of the summer rains requires a pilgrimage to the desert.

The summer rains are traditionally a time of renewal in this region. The Tohono O'odham, the desert people who have lived in this area for centuries, grow their summer crops using only the water collected from the infrequent runoff of the summer showers. The desert wildlife and plants depend on the summer rains as well, and to be in the desert just after a big storm is to witness one of nature's true miracles. The desert springs back to life after the brutal heat of June.

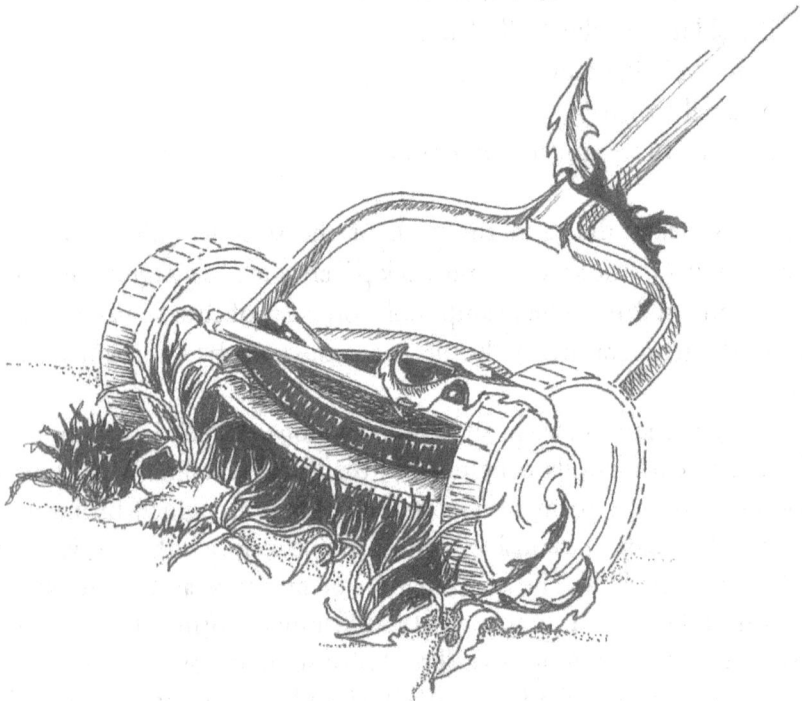

The Accursed Push Mower. (Illustration by Jan York)

The rains generally begin during the third week of July and a couple of weeks earlier in the high country. During the peak of the season, which usually occurs near the end of August, a rain every three days is not uncommon. Long-time residents of Tucson have learned to keep an eye on the sky over the Santa Catalinas, the big rugged range that dominates the city's northern skyline. May and June are dry months in the Sonoran Desert, and after weeks and weeks of intense heat with the relative humidity hovering in the teens, the local citizens start consulting the sky more than their wristwatches. If a big thunderstorm is brewing, the clouds over the mountains will begin to build around nine in the morning. By early afternoon the small puffy clouds of morning will have built into towering anvil-topped monsters, and the storms will start to slide off the slopes into the city by mid to late afternoon.

Accompanied by the distant growling of thunder, the first phase of the storm is a short blast of hot wind, full of dirt and other debris, as the storm shoves the stagnant air out of the Tucson Basin. The hot air is soon replaced by a fast-moving mass of cold, moist air dropping down from the storm cell. The visceral delight in experiencing the rapid transition from 105º to 75º brings people out from their homes and offices in droves. Television, movie theaters and work forgotten, people sit on porches and stand in doorways to enjoy the cool air and to watch the unmatched lightning displays as the storms roll across the city. If you are in the path of the storm, lashing winds precede a tree-shaking downpour that can drop an inch or more of rain in a fraction of an hour, punctuated throughout by the flash and roar of lightning and thunder. After the storms have moved on, the air over the city remains cool for the remainder of the evening. The first big rain of the season washes the dry heat right down the Santa Cruz River.

The next morning, I leave Tucson well before sunup. The storms have dissipated during the cool night and the sky is perfectly clear. The few feeble stars visible from the city multiply and brighten as the truck rolls east on I-10.

With the radio on, I sip my convenience-store coffee. There is nothing quite like the convenience-store coffee offered in the United States of America in the late 1980s. The low- grade beans are blended beautifully with the finest in Columbian paint thinner. One doesn't actually sip the stuff; it is not to be savored, but consumed. Quickly. My technique is an abrupt jerking motion with the head, keeping a firm grip on the cheap foam cup as the evil brew slams down the gullet. As a bonus, the ill-fitting lids have a tendency to send a stream of boiling liquid onto your lap. It may not be a gourmet experience, but it will damned sure wake you up.

With both windows rolled down, the cool, damp morning air howls through the cab. The economy stereo system is cranked up to maximum and the faint strains of an old blues song comes in over the radio: "When the levee breaks/I'll have no place to stay..." I have no idea that the song is a portent of the fearsome day to come. Oblivious to impending danger, I concentrate on fighting the classic motorist's battle: one twenty-ounce bucket of coffee, one sixteen-ounce bladder (and the bladder always wins).

A glutton for punishment, I stop for more mud in the small river bottom town of Benson. Different town and a different convenience-store chain, but it's the same coffee. The recipe must be nationwide. As I accelerate back onto the freeway running the truck through the gears, the cup of scalding coffee is perched insanely between my thighs. The sun is lighting up the eastern sky.

The orange sky brightens as I motor east up the long smooth grade out of the San Pedro Valley. The Little Dragoons are coming up fast on my left. They are not overly spectacular. There are no jagged crags, slashing canyons, or massive cliffs. The Little Dragoons are just smooth, crumpled hills. The lower slopes are thickly blanketed with mesquite, catclaw, ocotillo and the occasional saguaro. Rich in copper, zinc, and lead-bearing ores, these mountains have been mined on a relatively small scale for over a century. The highest peak in the range tops out at 6700 feet, which is not too impressive in these parts; the Little Dragoons are surrounded by massive sky island mountain ranges.

In some other more geologically challenged part of the country the Little Dragoons would probably be a National Monument. In Arizona (God's country!) they are just another range of desert hills.

For twenty years I have passed by these unassuming little mountains. I am coming here now because I have hiked and explored just about every other mountain range in the region. The big sky island ranges: the Chiricahuas, Rincons, and Huachucas, and the Baboquivaris, Santa Ritas and Santa Catalinas. The middling mountains: The Santa Teresas, the Mule Mountains, the Whetstones, and the Galiuros. The low, menacing desert fortress ranges: Superstition, Kofa, the Picacho Mountains, and the Peloncillos, Sawtooths and Eagletails. The lesser mountains: The Tortolitas and Sierritas, the Tumacacori and Patagonia Mountains, and the Winchesters and Tortillas. From atop Mount Lemmon, the nine thousand-foot summit of the Santa Catalina Mountains, there is no mountain range visible except the mountains on the big Tohono O'odham reservation that has not felt my feet. I like to walk.

Just west of Texas Canyon a beautiful mule deer buck browses in roadside salad on the freeway shoulder. He is so close to the road that the windblast off the tractor-trailer rig ahead of me whips down the tall grasses around him. Black muzzle dripping, he calmly ignores the scream of engines and tires. I consider stopping to give him a look through the binoculars but I have to give it up. At sixty-five I am moving too fast and there is another giant truck just behind me. The big buck is a good omen. One of my goals for the day is to spend some time looking for deer.

After passing through the fantastical geologic formations in the rocky trough of Texas Canyon, I go through the ancient ritual of missing the turnoff, studying the maps, missing the turn again, and finally, succeeding.

The smooth glide of tires on pavement turns into that low, satisfying rumble of a well-used pickup on a dirt road. Aaaah, yes. I skirt some mine workings, and relying more on my clearly reliable sense of direction than outdated USGS topology maps, work my way along the eastern slope of the mountains. The road soon turns to double-track, then double-rut.

At a fork in the road I park the truck and spread the maps out on the hood. The morning is clear and cool. The Little Dragoons are much prettier up close than when seen from a distance. The early morning sun highlights smooth, grassy lower slopes, which give way to oak forests on the higher elevations.

Retrieving my big 10x50 binoculars from the truck, I lean against the brush guard to survey the terrain. There is no sign of rain, not a cloud to be seen. To the south, the Dragoon Mountains lift from the grasslands, their fortress canyons providing safe haven for Cochise and his band of Chiricahua Apaches during the bloody Apache Wars, over a century ago. North and east, the country opens up into the vast range-and-basin landscape of eastern Arizona and western New Mexico. The golden morning lights up the sky islands and the lesser ranges: the smooth, grassy Winchesters, the rugged and remote Galiuros, the tall craggy Dos Cabezas, and in the distance the great blue wall of the Pinalenos.

A narrow track leads me westward into the mountains. The road follows the bottom of a canyon, twisting and turning pleasantly as it meanders up the slope. After about a mile, a barbed wire fence crosses the road. A "No Trespassing" sign hangs squarely in the center of the gate. Oh, well. Shifting into reverse, I see a small herd of mule deer on a rise just fifty yards to the south, a cluster of long necks, slender legs, and big ears, standing and staring. One by one, the deer break away and trot off to the south, with one smallish buck bounding and leaping behind the does. I leave the truck in the road, grab a canteen and camera, and take off in a hopeless attempt to get a good photo. The deer move steadily away (smart deer), and I get just the occasional glimpse of them through the brush. No photo opportunity here. The deer soon disappear over a distant ridge, and the hunt becomes a leisurely stroll down the canyon. When I return to the truck the peaks of the Little Dragoons are cloaked in cloud.

As I drive out of the canyon and continue north along the road that circles the mountains, the sky continues to blacken with startling speed. It soon begins to rain steady and strong. Wipers on, I peer through the windshield and creep along in low gear.

An hour later the rain has not let up and the desert is flooded. Losing the main road at a confusing intersection, I find myself on a safari track that leads north across the grasslands. With tall grass on all sides, visibility is reduced to a few yards, and I pretend I am in deepest Africa. Maybe I'll spook a rhino. Dropping off a low ridge I find myself in a newly made swamp, and quickly turn off the wipers for fear of having them ripped off by the thick mesquite branches that overhang the "road." Plowing through grass and mud marked faintly by the wheel tracks that I assume were made by the wagons of the first settlers, the truck suddenly lurches and spins out.

No problem. I shift into low range four-wheel drive. The truck doesn't budge. Hmmmmm. I roll down the window and lean out, stomping the pedal to the floor. The front tires don't spin. No four-wheel drive.

I pretend I am at home trimming the hedge.

Standing in the torrent and surveying the scene with an experienced eye, it is obvious that the situation does not look good. The shrubs ahead of and behind the truck are pathetically small; they are not nearly solid enough to anchor for a winch-out. Climbing back into the cab, I try backing up, and then attempt the old standby of rocking the vehicle out of the rut. Not a chance. The swamp is filling fast and the rain is not going to stop anytime soon. Digging the repair manual out from behind the seat, I sit in the cab and study up on the Chevy four-wheel drive system, circa 1984. Vacuum-actuated four-wheel drive system. Hmmmmm. I open the hood and finally find the vacuum line that has come off its fitting, thinking all the while about the long walk out in the rain. Joy of joys as the front axle engages with the familiar metallic clang, and with all four paws digging and flinging mud the little truck leaps out of the rut and barrels down the track. "Yes! Yes! Yes!" I scream, very happy for the wonders of four-wheel drive. Five minutes later I exit the track onto the main dirt road, which is rutted, rocky, and all of twenty feet wide. It looks like a freeway.

The repair manual ($9.95, good investment) goes back behind the seat, and I decide to circle the mountains to their western slope and do some walking. Most of the drive down the western side of the mountains follows a big watercourse, which makes me a bit nervous, but the rain finally lets up and the wash stays mostly dry. On I go, opening and closing about half a dozen gates on the way. This is cattle country: corrals, barbed wire, windmills, and a lot of very wet and miserable-looking cows.

A rough two-track road takes me up and out of the wash. The road ends at a windmill and cattle pond. The area around the windmill and pond is what is sometimes referred to as a cattle "sacrifice area," a wasted, trampled area devoid of all plant life save a few mature mesquite trees. Taking to the steep hills to the north, I walk around the cattle pond, stopping now and again to look through the binoculars. I am rewarded with views of cattle, cattle, and more cattle.

The desert has a weird tropical feel to it. The relative humidity must be close to one hundred percent. The mountains are bright green and the desert flora is in ecstasy; everything is fuzzy with new growth and exploding with life. The mesquite limbs are simply dripping water, soaking my back as I climb past and through them. The thorns are soggy and don't tear at my skin and clothes. It is suddenly seriously hot; the sun is beating down through the misty drizzle.

After less than an hour it becomes clear that this is not a good place to look for deer. The area around the cattle pond is so severely "cattle-bombed" that wildlife would obviously find better habitat elsewhere. Making my way back down into the valley, the rain picks up into a steady pounding roar, blotting out the sun, the hills, everything save what is right in front of me. It rains so hard my pants are covered with muck from the raindrops hitting the barren ground and hurling mud back up. The rain instantly stops just as I reach the truck. Very funny. Glasses fogged, totally soaked and overheated, I drive down to the big wash with the driver's door open, leaning out so as not to soak the seat.

As I drive up to a nice grove of mesquite, a dozen or so cows and calves scatter down the wash. The bull that is with them does not scatter, however, and as I park under a big tree I must carefully maneuver around the enormous animal, which is approximately the size of my truck. I stand near the cab until he slowly and rather insolently moves away. Keeping a careful eye on the bull, I strip, put on dry shorts, and spread my wet clothes out over the warm hood. Pulling the lawn chair out of the camper shell, it is time to relax and bury my hot feet in the cool wet sand. Eyes closed and melting into the chair, I sip lukewarm water from a canteen. Nice. Mr. Bull has rejoined his harem. Good. Giant animals have my respect.

I am nodding just on the edge of sleep when I hear a sound close behind me. It sounds like a very large animal snorting through very large nostrils. Glancing over my shoulder, I find myself staring into a pair of nostrils the diameter of twelve-gauge shotgun bores, the two of which are attached to the remarkably muscled body of Mr. Bull. I casually leap from the chair and move in a leisurely sprint for the truck. Once safely behind the tailgate, I turn and watch as the bull snorts again, spewing a pint of phlegm onto my lawn chair. I have obviously upset him somehow, probably for scaring the women and children. He walks very slowly between the chair and the truck. Moving to the other end of the vehicle, I cannot help but admire his attitude, his incredible strength. He circles slowly, and I in turn move around to the other side of the truck, wishing I had remembered to close the passenger-side door. He could rub it off without noticing. The cows and calves stand in a ragged line and watch as their hero routs the puny human. Totally humiliated, I can do nothing but stand and envy his huge balls, easily the size of cantaloupes. He gives me the Manly Eye. "Want some?" Mr. Bull is saying. I don't.

This incident serves to remind me just how much I hate cattle. Nothing really personal against the insect-brained, cloven-hoofed, shit-spewing, fly-adorned, range-stripping critters, it's just that they ruin the countryside.

Cattle have drastically changed much of the Western landscape. This part of the country was mostly grassland one hundred and fifty years ago. Today a great deal of the land is scrub brush, cactus, and cut bank.

As a boy growing up in southern Arizona, I spent a great deal of my youth outdoors, and it seemed as though I was always scrambling up eroded slopes and pressing through catclaw thickets. The open spaces and lack of people spoke to me of wildness and wilderness. Until I was twelve years old, I thought the land I explored was the natural landscape. Cattle were an ordinary part of the scenery to me. Once there had been buffalo and now there were cattle, brought into Arizona by the great cattle drives of the late 1800s. As a boy, I imagined Apaches and those who lived here before the Apaches came, all scrambling up eroded slopes and pressing through catclaw thickets.

When I was twelve my family and I went to a famous museum located near Tucson, and I saw an exhibit that graphically showed the effects of the introduction of cattle upon the Sonoran Desert and the arid grasslands of southern Arizona. The first model was a depiction of the countryside in the mid-1800s. The final model, one of four, showed the current landscape circa late 1960s. The four models revealed a gradual and terrible one hundred years of destruction. The great seas of grass had been stripped bare and the deserts reduced to overgrazed wastelands.

I was shocked. Our beloved Arizona has in effect been turned into a giant corral. The mythical cowboy, the glorified hero of my dreams, had brought in too many cattle for the arid land to support. The legend of the cowboy was never quite so sweet to me after that day.

In the year of 1850 there were few cattle in the American West. Vigorous resistance by the Apaches and other warlike tribes had slowed the advance of the U.S. frontier in what were to become Texas and the Southwest states for decades. With the end of the Civil War, the U.S. military became much better organized in its efforts to crush the native tribes. By the 1870s the Indian Wars were largely over, and Native Americans all over the West had been forced onto reservations. With native peoples and the bison eliminated as competition for the land, the cattle population in the West boomed; in 1884 it was estimated at 35 to 40 million. The effect of all those big, non-native animals pushed onto the dry lands was almost immediately catastrophic, but it is the longevity of the cattle industry's hold on the public land that is the poison. The rancher and his cow have a chokehold on the West and they seem determined to maintain their grip.

Take a drive down a back road across our public land nearly anywhere in the West. Park your vehicle. Take a walk away from the road; walk as far as you can in a day. You might see a few deer, or a herd of elk if you are in the high country, probably several species of small game and many species of birds, but there is one animal you can definitely count on seeing. Cattle. Everywhere. And isn't it thrilling? There is nothing like the sight of a clumsy bovine, a segment of cholla cactus stuck to its nose and barreling through the prickly pear, to get one's blood going.

I'm just being facetious. None of us journey into the back county in hope of seeing a cow. We would like to see a coyote, a mountain lion, or perhaps a bear, but most likely (except for the indomitable coyote) you won't spot a predator. All these animals are routinely hunted and killed by ranchers and their admirers in various government agencies. And it's not only the predators that are thinned out. Competitive species such as antelope, elk, and bison are now either nearly extirpated or their numbers reduced to mere fragments of their former numbers.

And that isn't the only part of the raw deal the public has gotten at the hands of the politically influential cattle industry. The list is long: We could have running streams bordered with lush riparian growth. We could have seas of tall native grasses. We could have beautiful, healthy deserts. We could have productive watersheds, but instead we have overgrazed, trampled, and eroded uplands that are far less fruitful than before the coming of the Cow. We could have a delightful natural environment to enjoy, but instead we have somebody's damned cows!

It not being economically feasible to raise cattle in many areas of the arid West (Arizona, as an example, produces a tiny fraction of the nation's beef supply), much of the Western cattle industry has to be heavily subsidized. Public lands ranching is just that; these folks aren't running their cattle on their land; they are running their cattle on somebody else's land. It's been called, accurately, welfare ranching.

The high demand for beef and leather is indisputable. And I love a fine beefsteak. Leather gloves are a joy, and a good thick leather belt like the one I am wearing today holds my ragged pants up with style. This doesn't make me a hypocrite. The vast majority of American beef, and all its by-products, like leather, are raised in the Midwest and Eastern states. You know, in places where it actually rains.

The Western cattle industry's grip, however, is finally beginning to weaken, as other hands now pull at the reins of the runaway horse. Although the vast majority of the American people still don't realize the damage done to their Western lands by the cattle industry, folks are beginning to catch on. It has been slow, because the truth seems to be always just over the horizon, just out of sight.

Popular environmental causes have many advocates. Certain aspects of our environment simply wring a powerful emotional response from people. Whales are good. People who kill whales are bad. Mighty mountains are to be cherished. Those who would disfigure them need to find different ways of making money. Wild rivers are precious, just try and build a new dam these days.

The tired, overgrazed deserts and mountainsides just don't get that same response.

Cattle have been a part of the landscape for so long that most of us take their presence for granted. Our damaged public lands are also part of our "normal" view. The range has been in such poor condition for so many years that most of us, myself included, find it difficult to imagine what a healthy Arizona would even look like. Grasses and shrubs haven't matched up well with whales and forests, and looming large in the imagination is the fairy tale of the earthy and magnificent cowboy, our steadfast steward of the rangeland.

Times have changed. The myth of the American cowboy can no longer carry a terribly destructive and minimally productive industry on his back any longer. When the cattle are gone from Western public range, we will all be witness to an incredible ecological recovery.

§

After being menaced by the bull and giving up on rest and relaxation, I pack up and drive back up the wash with a vague plan about hiking into the interior of the mountains. Just as I close the first gate it begins to rain again. This is a serious rain; the kind of rain that you know deep in your sodden heart is here to stay. The sun may never shine again. It comes down so hard that the wipers can't keep up. I finally turn them off, letting the water sheet off the windshield. It occurs to me that the steady rains of the morning have saturated the normally dry desert soil and further rainfall will not soak in, but will instead run off and seek low ground. In steep desert terrain like this the water will seek low ground in a hurry, so I drive at maximum safe speed plus about ten miles per hour.

It hails and then rains again, and water begins to run down the wash. When I reach the area where I expect to find my escape route, the wash is a quarter of a mile wide and filled from side to side with water. Somebody has done some quarrying here in the past, creating a low basin in the watercourse. The morning rains have created a small lake. Slipping and sliding through the increasing flow of muddy water, I retreat down the wash and finally find a road that takes me up and out of the watercourse. It is a big relief to be up on high ground.

The road I am on is a crude track. People that appeared to have been in a hurry have carved it out of the hillsides. The earth removed from the path of the roadbed is piled several feet high on either side and the effect is like driving down a chute.

During my wild run up the wash, the right-side windshield wiper became fouled with mud and other debris. Badly in need of replacement anyway (it never rains in Arizona), the wiper isn't taking any rain off the windshield, but is instead spreading a muddy streak across the glass and making an irritating squeaking noise. Leaving the engine running, I stop the truck in the middle of the road and fetch tools from the box in the camper shell. It is still raining hard, but I could care less. Being completely wet anyway, I relax and am soon engrossed in the repair job; removing the fouled wiper, cleaning it, reattaching it to the wiper arm, testing its operation, and then still dissatisfied, repeating the process. The rubber blade has nearly been torn from its groove and I decide to reattach it. The repair job goes to Phase 2: music required. I fiddle with the knobs and find an AM country station coming from some cow town out on the forsaken prairie.

Going around the hood to retrieve another tool, I happen to glance up and see a low wall of water rushing down on me. In an instant I realize that the road is an artificial watercourse and a trap for my truck. The high banks neatly gutter the water and leave me no escape route. Even in four-wheel drive the little truck has no chance of climbing the steep, loosely packed earthen walls. I jump in and slam the transfer case lever down into four-wheel drive high range, throw the transmission into gear, rev the engine, and dump the clutch. Going back down the road into the wash is out of the question; riding a flash flood into a lake might be briefly exciting, but would surely end in disaster. My only hope is to find an escape route. Actually, I have two hopes, the second being that the water won't get deep! I run through the gears as fast as I can, and as the truck plows forward into the rapid, muddy torrent, the water gets deeper fast. Six inches. A foot. I downshift to third and hammer the accelerator all the way to the floor. Wishing desperately for a big, powerful engine instead of the demon-spawned, tiny-pistoned, godforsaken, pathetic little six-cylinder I chose to get better gas mileage, I am forced to immediately downshift to second. Eighteen inches of water, and the truck is straining hard; any second I expect the howling engine to gulp some water and die. There is no break in the wall on either side as far as I can see. The flood is two feet deep, and hidden debris is banging into the front bumper and rattling its way back along the undercarriage. I slap it down into first, waiting for the engine to die as water pours up and over the hood. Sorry, little truck, but I think daddy is gonna have to bail out.

The flood is coming down the road very fast, and the pickup is soon going to be swept downstream. I desperately try to save the truck by flipping her up on the steep bank; if I can get two tires up and over the top of the wall she might stick, and I can crawl out and get her later when the flood subsides. Or the truck might roll, I realize, as the off-road tires bite and sling mud and the truck claws up the bank at an alarming angle, and then I lose control and dive back into the stream. Cold, muddy water pours through the open driver's side window as I keep the truck headed across the stream to maintain momentum. The truck shoots up the far bank, and this time I manage to hold it there for a few yards, and then suddenly the bank just ends. A road leading south and steeply uphill cuts through the berm, and I spin the wheel to the right, all the way to the stops. With all four tires on hard rock and the engine turning about five thousand rpm, the truck shoots up the little track like she's been fired from a rifle, with the fat tires screaming and smoking on the wet rock.

As soon as the road levels out, I leave the truck running and walk back down to the edge of the chute on rubber legs. Sitting on top of the bank with my legs sprawled in the mud, breathing hard and doing a bit of that high-adrenalin trembling, I am astonished that the pickup made it out of there. The foamy, muddy water, carrying tree branches, boards, and logs, rushes by with an impressive rumble and roar. Compared to Lava Falls on the Colorado this is a leaky faucet, but considering that a few short minutes ago it was a quiet country road, not bad. Not bad at all. It damned near nailed me, the wily Desert Rat.

The way forward and back (all routes home) temporarily closed, the wily Desert Rat follows the only path available; this most wondrous, magical, blessed road that saved my truck (and quite possibly my life). The road winds up a ridge and across a large, gently sloping plain to the base of the mountains.

The wiper remains disassembled, the repair job incomplete. The little knob on the stem scrapes its way up and down the windshield. To hell with it. The one in front of me works. The country station fades away to static as I pass behind low, grassy foothills. As I travel, I thank the road many times. Nice road. Good road.

I park the truck at the rim of a small, narrow gorge. The rain has shifted from downpour to a lukewarm mist. Soaked to the skin and burning with adrenaline, I strut around with my shirt off, alternating between gloating over my astounding driving skills and feeling like an idiot for being caught in a flash flood. The Gloating Fool. Lunch is a roast beef sandwich, fruit, and some very hot Serrano chili peppers, which leave me pouring sweat in the warm, steady drizzle.

While eating, I study the USGS topographical map of the area, which shows a small group of buildings in the center of what appears to be an interesting valley located just over the line of ragged hills to the north.

A word here on map reading. The U.S. government, not known for producing miracles, has nonetheless formed and somehow not abolished the wonderful United States Geological Survey. This agency no doubt performs a variety of useful tasks, but the production of their topographical maps is, in my mind, the most important. For a small fee one can purchase a middling-sized piece of paper that is packed with practical and accurate information. For instance, from where I sit the little valley to the north is invisible. The hills completely conceal it from my view. When I study the map, however, the complex jumble of elevation lines creates the valley in my mind. I can close my eyes and picture the area: high, smooth mountains to the west, lower but more rugged country to the south, and the valleys' natural east exit out onto the open desert.

Being able to create this mental image takes a bit of practice. After a few thousand trial-and-error entrances and exits from canyon, valley, and chasm, and a few thousand descents from range and mesa, all the while more or less continuously consulting a map, one gets a feel for what the little squiggly lines really mean. To the true enthusiast, map reading can be nearly as much fun as actually walking the land.

The buildings are probably occupied, but hopefully not. I load and strap on my pack, taking a last quick look at the map before folding it and stowing it away in a leg pocket of my camouflage pants (my favorite color is... camouflage).

Skidding and slipping my way down the near slope of the gorge, the sun unexpectedly breaks through the clouds. Wet stone is instantly transformed into glistening sculpture. The cliffs on the buttes ahead invite me into a hanging, twisting canyon. A glimpse of a deer through the tangle of oak in the canyon bottom: whitetail? The sound of a large, leaping animal: Whump! Whump! Whump! Even though I can't see it, it must be a mule deer, bounding uphill using all four legs at once.

The rocky channel of the canyon is roaring with water. It runs clean and clear, the canyon already swept free of debris by the morning flood. Leaping, tinkling, bubbling, swirling, falling, pooling, the laughing water swirls around rocks and tumbles over tiny waterfalls. Yesterday I could have casually strolled up the center of the parched watercourse, pausing perhaps only to negotiate a felled tree or a pile of rock. Today is a different world.

I finally tire of trying to keep my feet dry and take to the steep slopes, switch-backing up the broad back of a ridge that peaks into a shattered crest. Hoping for a look down into the valley, I stay on the crest of the ridgeline while working south. This becomes an ankle-twisting ordeal and I am soon lathered and winded. The rock on the ridgetop is jumbled and badly weathered. Walking on top, I can often see fifteen or twenty feet straight down through cracks and fissures in the rock. One misstep and a hard fall is waiting. Stopping to rest, I spot the obviously abandoned ranch house and outbuildings in the valley below. A nice, smooth ridge takes me neatly down into the valley.

The old ranch is a classic southwestern spread. The bunkhouse is fifty yards east of the main house, and a shed and workshop are to the west of the house and slightly uphill. The valley is very pretty. The little basin is nearly circular and uniquely symmetrical in appearance. It is a natural place to settle, and the weathered old buildings look like they belong to the valley. There are many healthy trees around the place. The ranch sits in the middle of a small mesquite flat and cedar and oak grow on the higher mountains to the west. The main canyon that drains the valley has been dammed, creating a haven for cottonwood and sycamore trees.

Shrugging out of my pack, I excitedly begin to explore the old buildings. Poking through the ruins, stooping now and again to pick up an old can or a piece of a wrought-iron bed frame, I can't help but wonder what it must have been like to live here back then. The ranch was probably built in the early part of the 20th century, perhaps the nineteen-twenties, and abandoned thirty or forty years ago. It may well have been built upon the ruins of an older ranch, and that on the ruins of Native American dwellings. The valley has a hospitable feel.

Excitement turns to melancholy as I prowl through the remains of the bunkhouse. All this work, all the hard, backbreaking labor that went into the creation of this place eventually went for naught. Ruins have a mystery to them, especially the really old ones, the stone and mud houses tucked away high on a canyon wall, but there is also something heartbreaking about them as well. I am exploring the relics of desert dreams.

The sun through the clouds creates a gray, surreal glow, which matches the ruins and my mood. I enter the ranch house last. It sits dark and somber, dripping water from its ruined eaves.

My grandmother probably lived in places much like this one. Some of her earliest memories were of growing up on a ranch on the Gila River near Redrock, New Mexico, just a few years after the turn of the 20th century. They ran a few head of cattle, and her father raised racehorses. When her mother died, her father took his two young daughters to Arizona in a covered wagon. The journey took them several weeks. Today the drive would take as many hours.

Once in Arizona, the two girls were raised on farms and ranches, with my great-grandfather supplementing his income with work in mining and timber cutting. Arizona was not yet a state and in most respects was still the frontier. Many of my grandmother's stories of her youth revolved around livestock; they always had chickens, pigs, cattle, and of course horses. Livestock was money in the bank in those days.

When people first tackled the frontier, they had to bring livestock with them in order to feed themselves. The infrastructure of the country was very primitive by modern standards, especially in the West. There was no electricity, no interstate freeway system, no airplanes, no trucking industry, and no railroads. Buying trucked-in produce was not an option, nor trucked-in beef. What you ate you and your neighbors grew yourself.

My father was the first in his family to earn a college degree. He was never tempted to leave his white-collar job and return to the ranch. Growing up on a ranch during the Great Depression doesn't exactly instill a sense of longing for the good old days. Paying jobs were scarce, and when they came, they usually didn't last long, so the range had as many cattle shoved onto it as could be, plus a few more thrown in for good measure. Hunting for deer and elk was not sport. It was a grim, businesslike affair meant to put food on the table.

When the country entered World War II thousands of good-paying jobs became available on the West Coast. My grandmother and her family left Arizona for California, and in the process traded subsistence living for a more comfortable urban lifestyle. When they came back to Arizona it was not to raise cattle or vegetables.

Is that what happened here? As the infrastructure of the country improved and the boom years of the war repaired the damaged economy, the country's population was increasingly shifting from a rural to an urban base. More and more people found that they had other options besides working hard seven days a week. With the price of beef uncertain and the rains even more so, the small desert rancher was always on the edge of disaster. Perhaps the lean years had so depleted the range that the land had become unprofitable.

As I roam through the shattered house, oblivious to anything other than my thoughts, huge masses of black cloud are pouring silently down the slopes of the main massif of the Little Dragoon Mountains.

The house is dark and dreary, most of the neat stuff has been taken, and I decide to investigate the bunkhouse one more time before hiking out of the valley.

Passing between the buildings, the view to the west brings me up short, and I can do nothing but stand and stare. It looks like a solid wall of water pouring down the mountain. If there was time, I suppose I would get depressed. The weather had seemed rather impressive, with the torrential downpours and flash floods and the like, but it has actually been just feinting and circling all day. Saving itself, it turns out, for the main assault.

The storm hits the old ranch with a howling burst of wind, and I quickly take shelter in the ruined bunkhouse. Crouching down against the west wall, I pull my hat down low and grimly decide to ride it out. My steely resolve lasts approximately two minutes, and I run back for the main house keeping my hands up under my hat brim to protect them from the slashing hail. Inside, I scramble over the remains of wrecked walls, furniture, and sections of roof, looking for a dry place. The roof is a sieve, managing only to collect and direct gallons of icy water streaming down onto my head.

The wind whips the broken house into frenzy. It shrieks and moans like the ghosts of the valley have come alive. Sections of the sheet steel roof tear free and cartwheel across the valley. It eventually occurs to me that the house could quite conceivably collapse, and I finally retreat to the only solid section of roof, which happens to be the small front porch.

Rain and hail take turns pounding the valley, coming in alternating, numbing waves. Wrapped in my poncho, I crouch on the porch for hours waiting for the storm to ease. The storm just does not let up. The black wall of cloud passes over, and is replaced by another, and another. Water pours off the shingles of the porch, which sheds water admirably. I commend its builder.

The Soaked Hiker. (Illustration by Jan York)

Late in the afternoon I decide I can't wait any longer. Pulling the poncho a little tighter around myself, I say a silent farewell to the valley, the shattered buildings, and the soggy ghosts, and trudge eastward down the only road. I feel like poor little Frodo Baggins, Tolkien's mythical traveler; small, wet, cold, and surrounded by forces much more powerful than myself.

The rain and hail continue like a giant bucket has been overturned high overhead. Lightning batters the ridgeline to the south, and I am grateful for the deep narrow canyon that the road runs through. After a mile of splashy walking, the lightning attack moves off to the south, looking for some other poor critter to fry. The canyon opens up and the sky to the east is clearing. Hope stirs in my sodden heart. After a few more hundred yards I can finally see around the steep flank of the ridge toward my truck, which is hidden from view down on the flat.

There is a fat green cloud rolling fast across the open grassland to the north. The hue makes me uneasy. Gray is fine, and white is wonderful, but clouds should not be green.

"Crrrraaaaack!" The flash is nearly simultaneous with the roar of the bolt. I lie for a few minutes in the roadside ditch, cringing as the barrage moves past, feeling a twinge of self-pity. Several twinges, actually, as cold water fills my pantaloons.

The road tops a rise and the climax of a magical, frightening, wonderful day is right in front of me. Tornadoes to the north! Twin funnel clouds are dropping from the wild green cloud, their mouths out of sight beyond the ridgeline. The tornadoes are fairly close, but protected as I am by the hills, I am not immediately worried for myself. My long-suffering steed, however, sits vulnerable out on the prairie.

After waiting until the tornadoes pass to the east, I make due north for the truck, enduring a nasty hailstorm that hits just as I crest the low saddle that leads from the canyon to the flatlands.

My truck is sitting just as I left it, unscathed by the tornadoes. I half expected to find it upside down in the bottom of the canyon. A half-mile north, however, I pass by a huge swath of flattened grass as wide as a football field and five times as long.

That's it. No more. I give up. I had planned to camp overnight, but this is getting ridiculous.

Time to get home and trim that hedge.

Twin Tornadoes. (Illustration by Jan York)

Bibliography

Nabhan, Gary Paul. *The Desert Smells Like Rain*. North Point Press, San Francisco, 1982.

Halka, Chronic. *Roadside Geology of Arizona*. Mountain Press Publishing, Missoula, Montana, 1983.

Jacobs, Lynn. *Waste of the West: Public Lands Ranching*. Privately published, Tucson, 1991.

Cook, Rob. "Ranking of States With The Most Beef Cattle." Beef2Live, 19 May 2021. https://beef2live.com.

Rite of Passage

Six inches of fresh snow lends grace to a cold autumn day.

I am fourteen years of age. Old enough to hunt big game alone for the first time in my life. My father, Greg and I sit around the tiny table in Greg's camper as a sharp, icy wind buffets the truck.

Sipping on a mug of steaming coffee, Dad explains the strategy of the hunt. As he talks, he strops his old hunting knife on a tattered strip of leather wrapped around a box of rifle ammunition. Bearded, weathered faces stare at me across the table as I nervously roll a .30-.30 cartridge around in my hands. Back and forth, back and forth, picking up the warmth from my hands, the smooth, simple deadliness of the loaded round is comforting. The powder packed behind the bullet complements the explosive mix of emotion in my chest.

With hands wrapped around a chipped, cracked mug and his battered hat crimped low, Greg adds sage advice to my father's rumbling sermon. The fresh snow, he explains, works in the hunter's favor, since any movement of animals leaves its mark in the snow. The two veteran hunters describe how patterns of animal movement can be determined from tracks. The advantage of fresh snow for the hunter is threefold: all animal sign is recent, easy to read, and the soft wet snow makes for quiet movement. "Only today!" Dad emphasizes. With the cloud cover gone and the temperature plunging well below freezing in the wake of the storm, the snow will wear a noisy crust tomorrow morning.

"And," Greg adds, "if you shoot a deer, and you don't hit him square, the snow makes for easy trackin'."

A grim thought. In all my hunting dreams, I have only pictured my first deer going down quickly and cleanly. The thought of a bloody trail in the snow makes my stomach jump, and again I feel the doubts of inexperience.

After the men finish their coffee, we go outside. The mesa looms to the south, dark wet cliffs and white slopes. At nearly 7,000 feet in elevation, the mesa is several hundred feet above the surrounding landscape. In the occasional places where the thick forest clears, sweeping views open to the south and east. Rising from a vast sea of pinyon-juniper, Arizona's unique "small forest," Mesa Redonda sits in a distinctive geological transition zone. To the south are the high volcanic slopes of Arizona's White Mountains, and stretching to the north for hundreds of miles is the sedimentary wonderland of the Colorado Plateau.

Nervous and apprehensive about the upcoming hunt, I fidget with my short carbine, swinging it up time and again to find the sight picture against a patch of snow, a tree, a boulder. What if I don't see the buck? What if I miss him? And what if I wound him, and he runs and runs and I can't find him? The frantic wind puts me on edge. In contrast, Greg and my father are portraits in manly, experienced nonchalance. The cold wind seems to bother them not at all. I envy the casual, practiced way they handle their weapons, their broad deep shoulders, and their big strong hands. They hold their rifles like they were born with the things in their arms.

Before we set out in the early afternoon, Dad gives me some final advice. He grabs me by the shoulder and gives me a gentle shake. It is a rare gesture. "Son," he says, "you're a crack shot. I wish I could shoot like you do. Hell, I wish I could take credit for teaching you. I can't. You're a natural with a rifle or shotgun. There are a lot of deer around here, and we are sure as hell the only hunters crazy enough to be out here just after a blizzard! Greg and I will work the brush, and anything we kick up will likely come your way. You'll do fine. Just keep your head up and your eyes open, and do what comes naturally!" He tilts his head back and laughs, and I laugh with him. Suddenly, I can't wait to hunt.

The tactics we follow are not mysterious and complex. Greg goes one way, Dad goes another, and I hunt the base of the big mesa. The senior hunters have granted me the prime area. They will hunt the difficult cedar breaks to the south of the mesa.

The snow is laced with deer tracks. They lead here and there, but mostly north and south, winding between the trees. The deer seem, by their tracks, to know where they are going; where the terrain allows there is a straight-ahead determination to the paths. The dark mesa is close on my right hand after the first steep ridge. The cliffs block the wind, and here the pine boughs are bent down under the weight of the wet snow. On the open ridges and down in the flats, the tree branches have been whipped clean.

The lee of the mesa is very quiet compared to the howling racket of the ridgeline. The snow muffles my footsteps. Excited and breathing hard, I am expecting to see a huge, leaping buck any second. Cringing over any small noise I make, I move upwind through the forest. Time and again, wet snow cascades down on my head when I disturb a small tree or branch. Once I have to stop and clear the barrel of my rifle after a deluge, and after that I am careful to keep the muzzle down when I pass under trees.

Moving slowly and carefully as the afternoon creeps on, stopping often to watch and listen, I see no deer, elk, or any other animal. The clean blanket of snow, however, is tantalizingly embroidered with the tracks of all manner of critters.

By late afternoon the forest is in deep shadow under a churning, overcast sky as the tail of the big storm clears the mountains. I decide to turn back when the snow begins to freeze. As I make my way back, disappointment sets in. The wind is now squarely on my back, pushing my scent far ahead, and I am making too much noise. My tired feet are making irritating crunching sounds instead of the pleasing, sneaky shuffle I have been maintaining for hours. I will surely not see any deer now.

But I want to. I want to see that leaping buck. I want to hear the sharp supersonic report of my rifle, smell the gun smoke, and watch the deer go down. I want to complete this ancient rite of passage. More than anything, I want to please my father.

Yearning to see a buck standing in the tracks he just made, the trunks of the jack pines start to look like deer legs, and the old fluttering dead oak leaves start to look like deer's ears. The shadows are full of mystery, of deer barely seen, and I creep along peering through the gloom, fourteen years old, tired and hungry. The rifle is like ice in my hands. The brooding mesa depresses me and the roar of the rising wind in the pines makes it impossible to hear. The wind whips ice into my face, making my eyes sting and water. Dejectedly, I mentally give up and trudge along with my head down, lost in thoughts of failure. What am I going to tell my father? I have seen not a single deer.

A noise at the edge of my consciousness brings my head up, and there he is. A beautiful young buck standing squarely in the center of a little clearing. With instincts honed through hours of bird hunting, trap shooting, and target practice, the little carbine sweeps up. As the rifle comes up, the hammer comes back, and my right cheek settles firmly against the stock. The sight picture is automatic: the lower chest of the deer just behind the joint of the foreleg sits on top of the front sight, the outline of which is nestled between the ears of the rear sight. The buck is less than thirty yards away. With this same rifle I can more often than not knock an empty .22 cartridge box from a tree stump at fifty paces.

He's mine. I've got him.

I can hear the deer breathing. He has been running hard. His breath comes in short, explosive bursts of vapor from his black muzzle. He is looking right at me.

Everything stops.

I can hear every sound in the dark forest. Each detail of the deer and the forest beyond is amplified with splendid clarity. The wind moves the pine branches. Snow and ice slide from the branches to land softly in the snow. The dark eyes gleam wildly in the waning light of late afternoon as he pants, chest heaving. I look him in the eye, and the moment that every boy who lies awake at night, dreaming of his first deer, is gone. And for me, gone forever.

Heart pounding, mouth open wide, I try to breathe without sound. I lower the barrel of the rifle very slowly, enough to get a clear view of the deer. The fragile-looking forelegs are covered in mud and snow. A tiny trickle of blood runs down the left front leg. The slender legs flare into hard muscle at the shoulders and flanks; masses of trembling stone. Chunks of snow, black mud, and tiny shards of ice sprinkle his back. The powerful, thick neck narrows gracefully to support the small head. He has an even, perfect rack, three points to a side, not counting the eye guards. His coat is damp, and the power of the packed muscle beneath it is obvious as he shifts slightly, prancing with the tiny dancer's feet.

The buck suddenly tenses, the muscles along his body bulging and clenching as the gorgeous head comes sharply up. My first deer is going to bolt! My finger starts to squeeze the cold trigger as I realign the sights on his heart. In horror, I snap the barrel up and back so quickly it slams me across the cheek, drawing blood and bringing instant tears to my eyes.

When my vision clears, the buck is gone. Vanished. A small cloud of vapor is whipped away rapidly by the wind. Letting out a long, slow, ragged breath, I carefully ease the hammer down.

That night, around the table in Greg's camper, we share our afternoon's experiences. Lots of sign, but nobody even saw a deer.

But Nobody Saw a Deer. (Illustration by Jan York)

§

The year after my deer hunt took place, "The Guns of Autumn" came out on CBS. Animal rights groups and other activists were challenging the ethical foundations of the hunting traditions while attempting to insert a new moral code regarding human treatment of animals. The debate on the ethics of hunting was howling like a hot wind through the country, leaving men like my father confused and angry, afraid of losing their enduring and cherished way of life.

For my father and many of his generation, hunting was part of the backdrop, an essential ingredient of living and one that had never, to his knowledge, been seriously contested. This new surge of opinion coming from across the entire nation, expressed with eloquence and undeniable passion, came to him as a wrenching shock. The skills and beliefs of the hunter were being held in open contempt.

The strong moral posturing of the anti-hunting and animal rights groups generated the majority of my father's anger. Their positions openly suggested that hunting was a pastime of the lower classes, a barbaric sport that appealed to only ignorant, brutish peasants. This savage use of typecasting, which seemed to intimate some sort of social class warfare, was extremely confusing to me at the time; my father and his hunting friends were all white-collar professionals with college educations.

Raised within my father's family traditions, feeding my parents and sister with what I perceived to be the bounty of the land was something I took great pride in. Due largely to my youthful efforts as a hunter of birds, small game, and fish, we always had wild meat in the freezer. It had never occurred to me that hunting could be wrong. Confronted by people who were horrified by the hunting of mammals, I found myself defending the tradition even after I discovered that I could never pull the trigger on a big game animal. Interweaved into the sentiments against hunting, pain and suffering of animals aside, was the dark symbol of the gun (and to a lesser degree, the bow and arrow). Take away the gun, the projectile-hurling weapon so often turned against mankind, and the arguments might shift. If deer were hunted with a net, or a barbed hook, would the rage against the hunt burn so brightly? I think not.

The issues of animal rights and environmentalism vs. hunting are obviously complex, generated by the intricacies of people's intellectual and emotional positions on these issues. Everyone has a line to draw in the sand, if you will, when it comes to the killing of living things for food and other products. That line, however, is not necessarily going to set the boundary on a tidy set of beliefs. People's emotional reactions to killing, death, and pain, applied against their preferences for food, lead them often on a tortuous ethical path. Even animal rights activists have their righteous battle lines placed at varying distances from mammals, fish, insects and other forms of life. To one man I know, the killing of wolves borders on sacrilege, while the killing of fish for the table is a practical matter. "Gotta eat something with eyes," he'll say, without a trace of guilt. He has made his peace, drawn his line in the sand.

Here is an example of just how convoluted and mysterious people's thinking can be on this subject: Picture a fine restaurant, a table laden with food, and hungry people engaged in earnest conversation over their meals. Two of the people at the table were speaking out very strongly against hunting (I found myself once again in the very familiar role of Defender of the Hunt). The first anti-hunter, who considers herself to be an environmentalist, told me, "If you were a real hunter, you would hunt with a spear! You would meet the animal on its own terms - not over the barrel of a rifle."

This woman makes a living in the Alaskan commercial fishing industry, and her resume includes the job of deckhand on fishing boats. The deckhand actually hauls the net, and the netted fish, out of the sea. She once participated in a protest against the killing of wolves (an animal in relatively rare supply) before hopping on a fishing boat to go out and kill fish by the ton.

The second woman at the table spoke out vehemently against deer hunting, saying it was cruel, unnecessary, and the result of "macho jerks wanting to prove themselves as men." This conversation took place while both these people were eagerly devouring large salmon filets!

This type of thinking is rife, and it can be taken to bizarre lengths. Many members of my family have considered themselves to be vegetarian for years, even though they eat fish. One could infer from this that fish are not animals but are instead some sort of swimming vegetables with eyes: "Salmon, beets of the sea!" One can be a vegetarian, evidently, simply by declaring what one chooses to eat to be a vegetable.

I have read about people who have taken strong positions on anti-hunting or animal rights and live a strong, discrete life in accordance with their beliefs. Speaking out against the killing of animals, they don't eat fish or other marine life. They also don't wear leather, or fur, and don't use any products extracted or manufactured from the bodies of dead animals. This is a position that commands respect. Unfortunately, I personally have never met one single anti-hunter or professed animal rights activist who leads such a discrete life.

"Fur is sacred, fins are food!" seems to be the unspoken rallying cry. Why is there this philosophical difference in the taking of mammals versus fish for food?

A Web search for anti-hunting groups quickly yielded a list of over 360 such organizations. A similar search for anti-fishing groups produced not one, and few of the animal rights people seem to take on sport or commercial fishing as part of their charter. The emphasis is: mammals, mammals, and mammals.

The symbol of the hook and net appears to charm the psyche in a similar manner as do the plow and the pitchfork. The classic profile of the fishing boat seems to float gently alongside the pastoral scene of tractor and barn, all tools of our peaceful farmers of field and sea. Appearances here are deceiving. The modern fishing boat is, in effect, a fully automatic weapon, equipped with sonar, radar, powerful engines, a skilled crew, and big nets. Fish are killed by these craft in such numbers that they cannot be counted individually; they must be tallied by the ton. The value of the individual animal is lost in the net - a net full of fish becomes comparable to a basketful of corn. Perhaps if deer were mowed down in equal numbers, the sheer volume of death would render it a harvest rather than a murder. One-on-one battles between the well-armed modern hunter and his prey bring forth an outrage that is strangely lacking against the industrial slaughter of fish.

The issue between mammals as food and fish as food is one of perceived value of life. The life of a marlin, for instance, appears to be of lesser worth than the life of a deer. This may be because fish live underwater, and are therefore far enough removed from our hairy, air-breathing existence for us to be readily able to identify with. Fish may also be perceived to be lower down on the evolutionary scale, and thus an animal more suited for food than mammals.

All of the above viewpoints play a part in people's thinking, but the real issue, I believe, is one of quantity. Fish are imagined to be in such abundance as to be of nearly infinite supply. Our bloody history has shown us that only when an animal is pursued to near-extinction does its significance rise. Grizzly bears, for instance, were once hunted like vermin because they were a professed threat to people and a definite threat to livestock. Today, with perhaps only a thousand of these bears surviving in the continental U.S., the grizzly is now a revered animal.

Bison, once blackening American plains with populations estimated at over 30,000,000, were pursued and killed with such vigor that by the late 1800s there were an estimated 1500 animals remaining. Only after this commercial bloodbath had ended was there a hue and cry for the demise of the mighty buffalo. When considering the mechanized efficiency of modern commercial fishing, I have little doubt that our oceanic fish will soon follow. Perhaps as soon as in my daughters' lifetimes one will be forced to visit a "fish preserve" in order to view the few remaining wild fish. And, of course, fish will have vaulted up the "value scale," and the few remaining predatory fish will rank right up there with the exalted grizzly bear.

When I gave up hunting a few years ago, I also stopped fishing. I personally cannot make the distinction in value of life between a fish and a mammal. Fish are truly beautiful animals, sleek and powerful and so suited for the environment they live in as to make the envy of bird or fish a difficult choice. Hauling an animal out of its home with a net to die of suffocation is no less cruel than a hot bullet to the chest.

The hunted ungulates such as deer and elk, traditional prey for the modern American hunter in the continental U.S., are managed by so many government agencies as to in effect turn them into domestic animals. Deer and elk live in the wild, or at least semi-wild, areas, living as they have for millennia, but their populations are closely monitored to ensure a viable population year after year. In contrast, the control of commercial fishing on the high seas is nearly ungoverned. The U.S. and other concerned nations try hard to limit and control fishing in their own waters, but fishing international waters is in effect a free-for-all, with giant trawlers mining the ocean floor with a merciless and destructive efficiency. The world's fisheries, the once-limitless breadbasket of the planet, are under such extreme strain that many concerned scientists feel they cannot hold out much longer. While human survival does not depend upon a thriving Colorado elk population, many people around the world depend upon the sea for sustenance. Our emotions, however, continue to lie with the mammals.

What is depressing about this age we live in is that not only are we making the same mistakes made in previous centuries, but also that the mistakes are being made by those who consider themselves enlightened. While those who exterminated the buffalo can be written off by environmentalists as little more than sawed-off-at-the-eyebrows primitives, and the killers of the grizzly and wolf as greedy pirates, today we have "right-thinking" individuals, people who try in every other aspect of their lives to tread as lightly as possible, killing wild animals for a living. The extermination of the bison may be forgiven as an unintended and unforeseen outcome, because the folks back then just didn't know any better, but by now we should damned well be much more progressive in our thinking and our actions. The cultural phenomenon of using the earth's resources to support a lifestyle, and not a life, continues nearly unchecked. Proper management of animal populations (including fish) to ensure a healthy and continuing, self-sustaining population, should be the overriding ethical boundary.

The price of failure will be high. Will people looking back a hundred and fifty years from now, living on their tired and depleted planet, forgive us because there was no actual plot to exterminate fish? Or will they damn us for being stupid?

BIBLIOGRAPHY

Halka, Chronic. *Roadside Geology of Arizona*. Mountain Press Publishing, Missoula, Montana, 1983.
National Bison Association. https://bisoncentral.com. Accessed April 5, 2000. Retrieved 4/5/2000.

The Company of Bears

"There's something big up there! Really big. It might be a moose."

Lisa crouches behind me as I bring the heavy 10x50 binoculars to my eyes.

"What is it? What is it? Can you make it out?" she whispers in my ear.

A large, hairy ass the width of a small pickup truck fills the lenses. The massive head turns to the side, brown muzzle dripping. It's a grizzly.

"It's a bear! I can't believe how big the damned thing is. Let's hide!"

With that noble comment, we quickly slide into the brush and peer out at the giant bear. The bear, sixty yards or so up the slope, is either unaware of our presence or completely ignoring us. It buries its head into the bushes again.

Lisa breathes into my ear, "It's eating berries!"

For the better part of an hour, we sit and watch the bear eat. It has an efficient and steady feeding style, methodically taking the miniature blueberry bushes into its mouth and stripping the branches through its teeth. The steady breeze from the west ripples the bear's long blonde fur across its broad back. While watching, we quietly take the ridiculous "bear bells" from our boots. The Park authorities had recommended these bells to us, small metal balls which rattle loudly as you walk and have the intended purpose of warning bears of the approach of hikers, thereby avoiding a surprise confrontation. The bells are silly, insufferable things, and we both felt foolish wearing them. Taking them off is a relief; now we can blend in with the landscape and move around without disturbing all the wildlife in the area.

"It's beautiful."

"It's incredible," I reply. It truly is. Spears of golden, slanting light are breaking through the cloud cover, and one falls on the magnificent bear like a divine spotlight, revealing the highlights in its long fur. The raw power of the animal is obvious. Its shoulders and back are packed with great mounds of muscle that dance and play as it feeds. Despite its size, the bear is obviously quick and agile, and clearly unconcerned about its own safety as it calmly grazes on berries. For one long period of several minutes, it never raises its head for a look around. There is nothing like being at the top of the food chain.

Before us, now on the crest of the ridge that forms the southern horizon and shining like a big hairy beacon, stands the golden bear. Beyond the bear, the tundra stretches without end, green and blue and misty gray into the fathomless distance. The immense emptiness of the Alaskan interior dwarfs all things, even the great mountains behind us, challenging the mind to grasp the sheer size of it. Despite the thousands of miles I have walked in country that I naively felt was wilderness, it is my first true sighting of a brown bear. The animal suits the landscape so impeccably; for a short time, my head swims and I lose my place, forgetting for a moment the road, the towns and cities beyond, the human world that is leaning so hard on places like this. We could be anywhere in time on this land; it is a primal view into the past. A humble perspective is not optional; I have waited all my life for this, but I can't admit the awe that threatens to burst my chest.

"It is definitely the dominant land animal. However, it is difficult to be afraid of something that's grazing," I whisper to Lisa. "If it was up there killing a moose or something, that would give me the willies. Right now, it looks like a big hairy cow."

Lisa is not fooled. She knows what I am feeling, and what this means to me. She replies, "Cows don't have big fangs and run around eating caribou. Besides, if you aren't afraid, then why are you hiding in the bushes? And whispering?"

"Good point. Why isn't it eating caribou?"

"Caribou run fast. The berries just sit there."

"Another good point. You're just full of wildlife lore, aren't you? If you're so knowledgeable, then why did we camp in the middle of a bear highway?"

"I didn't know it was a bear highway. I do now. Those tracks were a lot bigger than my little boot prints. Big tracks."

"That thing could be down here jumping up and down on our battered carcasses in about ten seconds. And you know what? There is nothing we could do about it."

"That's not true. We could wave our arms and say, "Nice bear." She is laughing, softly, but it is a rather strained laugh. "Have you noticed that there are absolutely no trees to climb around here?"

"Well, there is that one," I say. "You know, the one the porcupine climbed? That's the only one we've seen in three days."

"Maybe we should go camp by the tree, then. If a bear comes around, we could climb it."

"No, the mosquitoes in that canyon would kill us. Let's take our chances with the damned bears."

After finding the fresh bear track directly in front of our tent, we had convinced ourselves that moving our camp off the river bottom would be the prudent thing to do, but now we're not so sure. There is no place to camp up here, anyway. The brush is so dense there is no place to pitch a tent.

After the bear moves on, we make our way back down into the river canyon, move our camp as far away from the water as possible, and hope for the best. Being from Arizona, we are not accustomed to the company of bears. This is not the tame wilderness experience we have become so familiar with, exploring our semi-domesticated pockets of relatively untrammeled land that we call "Wilderness Areas." Black bears still roam the remote places, of course, but compared to the great grizzlies, black bears are shy, retiring creatures that customarily avoid humans. When hiking the backcountry in the Southwest you can move without fear of the wildlife, as human beings are the dominant animals (minding your step in certain areas; there is the occasional rattlesnake to step on). A rancher killed the last known surviving Arizona grizzly in 1935, and intense "predator control programs" doomed the wolf not long after. Without the great bears and the wolves, most of America has become an emptier place. Hiking through the more isolated areas of the Southwest is like looking at a striking painting with some of the necessary colors missing. Something magnificent, powerful and deadly is gone.

An omnivore (although mostly vegetarian), eater of berries, grasses, sedges, nuts, insects, and carrion, an opportunistic hunter and a creature eminently capable of defending itself, the brown bear was once common throughout the American West. Weighing up to 1,500 pounds, arguably the world's largest land predator (the polar bear is sometimes reputed to be larger), grizzlies were at one time widespread throughout an astounding array of habitats. There were brown bears living on the high plains east of the Rocky Mountains, along the coasts and the lush interior valleys of

California, and equally at home in the rugged, arid country of the Southwest or the frozen tundra of Alaska and Canada.

By the year of 1850, the grizzly was on its way out. As the tide of humanity rolled into the West, the grizzly, an estimated 50,000-100,000 strong before the coming of the Europeans, was eradicated from nearly every square mile of what was to become the United States. Long accustomed to a tame landscape, the Europeans looked at the vast North American wilderness with as much fear and trepidation as greed, and sought to dominate, domesticate, and exterminate with as much speed as could be mustered. The subjugation of the American West was short, savage, and ruthless. The grizzly, like the Indians, buffalo, and wolf, went the way of all things that stood in the path of progress.

Today perhaps 1,500 grizzly bears exist in the lower 48 states, living a tenuous existence on the few remaining chunks of wild land still able to support them. Six fractured populations of grizzlies are all that remain in the U.S., and most of those are found on or near the U.S.-Canadian border. The completely isolated Yellowstone population is the sixth. All of these grizzly habitats are under siege from the usual human pressures for more land, more roads and more fuel, pressing hard on the ancient fortress of the Northern Rockies. Grizzlies require very large amounts of space. Unlike the coyote, they cannot thrive on the edge of civilization; unlike the highly adaptable black bear, they cannot get by with a few square miles of forest or mountainside. Grizzlies need big, deep wilderness with exceptionally healthy ecosystems in order to survive.

It would be a major environmental triumph to reintroduce the grizzly to Arizona or New Mexico. Even a handful of grizzlies lurking in the Blue Range or the Gila Wilderness would not only lend a completely different aura to an outing in that region, but would give both states a measure of regard. And it would be more than prestige: To every resident, be they urban or rural, whether they ever venture away from paved environs or not, there would be the awareness that out there, something glorious had returned.

§

The sight of a bear can produce extraordinary effects on even the most composed of people, including increased heart rate, heavy perspiration, and a sudden urge to climb a tree. More advanced effects, usually found in people who are a bit more excitable by nature, include screaming, running, and even bursts of raw hysteria. The most extreme effect is something I call Bear Paranoia, which has a tendency to strike hardest at those who have highly developed imaginations, and has influence even when bears aren't around.

Bears and people have occupied the same planet for many thousands of years. Our deep primal memories include the knowledge that large, fierce bears were at times a terrible enemy of our rugged ancestors. In addition, the ancient memories of large, fierce bears undoubtedly include the knowledge that, in times of need, the tall frail things are an excellent source of proteins and other vital nutrients. On the other hand, bears instinctively know that the even though the tall frail things are weak, run with a pathetic lack of speed and have no natural weapons for any bear to fear, they are rather clever, easily provoked, and have a nasty tendency to be armed to the teeth. When bears and people come in contact, these combined memories can produce a certain tension.

The intellect, or the ability to think and reason, is one thing that separates us from other animals (including bears). What a lot of us don't like to admit is that our primal knowledge dramatically influences how we think. At the intellectual level, people think about bears; people analyze bears, some professionals and avid amateurs study bears, and a few go so far as to ponder the meaning of bearness. After years of inquiry, some of us come away with the clear conclusion that we are simply scared shitless of bears.

This is Bear Paranoia.

Before our current trip, my wife Lisa had already spent two summers in Alaska. She had hiked into the wilderness, of which Alaska abounds, and had a close and frightening encounter with a black bear. Two summers in Alaska is a long way from being an old sourdough, but at least she had been there. The problem with Lisa is that she is disappointingly practical. If confronted by a bear she would be concerned, but if no bears were in sight, she would not be particularly worried. The key in her mind is to not do anything stupid, like use a fresh salmon for a pillow. When Lisa and I first started planning our trip to Alaska, I naturally sought a man's viewpoint.

As luck would have it, the first two people I talked to were men that I worked with, fellow engineers, mature family men, and both heavily stricken with Bear Paranoia. You would never know it just to look at them. When I mentioned my Alaska plans to the first man (whom I will empathetically refer to as Conrad to conceal his identity), he got actually bug-eyed and immediately broke into a light sweat despite the cool office environment.

"I wouldn't go to Alaska if I were you!" he proclaimed. "Don't take your wife there, not hiking around in the bush. You'll regret it. The risk is too great. I would never go to Alaska myself."

"Why?" I asked him. "What's the risk?"

"Bears. Big bears!"

"What's to fear about bears?" I replied, slightly amused at his unusual display of emotion. "If you take reasonable precautions..."

"Reasonable precautions!" he interrupted, "There aren't any!"

Not understanding his agitation, I laughed and went about my work. Later in the day Conrad pulled me into his office, sat me down, and recounted an experience of his:

He and his family were visiting relatives in Idaho, and a group of them went for a stroll one day on a trail leading into a remote mountain area. Conrad, a well-conditioned man who always keeps himself in shape, ran on ahead of the group. When he had gone some distance, enjoying the pretty scenery and the fresh mountain air, he looked up and saw the outline of a bear's head peering at him from out of the thick brush along the trail. Knowing this wasn't grizzly country and assuming the black bear would flee into the forest at his approach, Conrad did not break stride. When he got to within about forty yards of the bear, the bear had not fled.

Conrad was studying Tae Kwon Do at the time, and the explosive burst from the lungs that students from his school were taught to make when striking an opponent was a very loud "Bah!"

"Bah!... Bah!" yelled Conrad.

The bear charged from the brush, right at him.

Conrad is certain he would have been killed if he hadn't been in excellent condition. As it was, it was a very close thing.

He turned and ran back down the trail, almost immediately realizing that the bear was not bluffing and that he was quickly losing the foot race. He picked out a large tree, a spruce or fir growing at the edge of the trail, and up he went, climbing hand over hand from branch to branch as fast as he could. The bear followed with no hesitation. Now in a state of complete terror, my friend continued to climb. The grunts of exertion from the pursuing bear and the sounds of large claws scrabbling for purchase on the trunk drove Conrad into a frenzy. He climbed faster still. It was a very tall tree, over a hundred feet high, and he pushed through the crown, squeezing through the thin branches. When the trunk narrowed to a thin spire of limp wood, he could go no further, and he simply hung on, exhausted, looking down between his feet at the climbing bear. The bear stopped when its head was a foot away from Conrad's feet, its massive limbs clamped tightly onto the slender trunk. They looked at each other, swaying back and forth at the very top of the tree. After an eternity of this, he heard voices coming from the trail as it switch backed down the slope. He yelled for help - voices answered! It was Conrad's family, meandering up the trail. The bear turned its head toward the voices, froze for a moment, and then simply released the trunk of the tree. In a controlled fall that broke thick branches all the way to the ground, it hurtled from the tree and bounded into the forest.

My friend has good reason to be afraid of bears. Certainly, he made a series of mistakes in his encounter with the bear. That is one thing I learned from his horrific experience. Don't yell at bears.

The other person I talked to about Alaska is a man with some Alaskan credentials. As a young man, Scott had gone to Alaska to seek adventure. He found it. Shortly after arriving in Alaska he found a job working as a prospector of sorts, seeking out old gold mining claims deep in the backcountry. The rise in gold prices and the development of mining technologies had made the working of long-abandoned claims sometimes very worthwhile. His employer would fly him into a remote area, and the sight of the disappearing plane would be Scott's last contact with the human race for days, or even weeks. Scott saw only one bear during his lonely days in the bush, a big grizzly, but he had many tales of the lush Alaskan landscape.

One day over lunch, he casually asked me what kind of rifle I would be taking to Alaska. I hesitantly replied that I had thought about taking my .357 magnum revolver, but then again maybe not. Gun laws being in the state that they were (and are) in this country, it wasn't a simple matter to take a gun from Arizona to Alaska. There was a certain amount of red tape and hassle to go through that I felt might not be worth the trouble.

Scott was shocked. First of all, going unarmed into the Alaskan bush was unthinkable, he maintained, and secondly, a .357 was not enough gun. The bottom line, he went on, is that a pistol doesn't have enough punch; after all, you're talking about a thousand pounds of enraged tooth and claw coming at you at twenty or thirty miles per hour. It takes a big slug to stop a big animal. He recommended a big rifle chambered along the lines of a .375 Weatherby or larger, or a .12-gauge shotgun loaded with rifled slugs.

It was during these lunchtime discussions, full of talk about things like weapons and ammunition, terminal effects and ballistics, fangs, claws and brute strength, that I began to get some insight into why there are no grizzly bears left in Arizona, and so few remaining in the entire continental

United States. People are afraid, very afraid, of grizzly bears. Without ever having had a close encounter with a bear, and never in my life having seen a grizzly, my head was nonetheless filled with scenes of huge slavering bears lying in wait, cross old bears ready to charge at the slightest provocation, and worried mother bears looking out for their cubs.

Even though I felt slightly ridiculous, I eventually yielded to this ancient terror and decided to buy a 12-gauge shotgun. A trip to the local gun store yielded a nice Remington Model 870, short-barreled and equipped with rifle sights. I bought a few boxes of magnum rifled slug ammunition, and Lisa and I made a trip out to the range to try out the new "bear gun."

The bear gun kicked like a field cannon; the recoil would consistently lift my forward foot off the ground when I fired it. After firing the shotgun several times, I found it to be sufficiently accurate at close range, and it certainly had an impressive amount of power. My wife fired it once. The recoil drove her 110 pounds back a few inches and pounded her shoulder black and blue. Lisa walked over, handed me the gun, and calmly told me that she would rather be mauled by a bear than ever fire that thing again.

We traveled to Alaska by train and plane, and the faithful bear gun went along, red-tagged and backed by the required paperwork. And I carried it. Once. After having my feet on the ground in Alaska for a couple of weeks, my Bear Paranoia disappeared. Through a little experience, and by reading everything I could find about brown bears and talking to the locals, it became clear that the risk of being attacked by a bear is very low. Brown bears almost never attack people. Confrontations almost always end peacefully. There are the rare exceptions, but to use automobile travel to illustrate a point, you stand a far better chance of getting into an automobile accident on Alaskan highways than in getting into a serious encounter with a grizzly bear. People have become desensitized to the grim toll of automobile fatalities. That you can die an ugly and violent death just going out for a box of doughnuts is something that has been completely accepted by the American people. Folks still (amazingly!) ride around with their seat belts off and their children scampering around inside the car. Grizzly bears, on the other hand, having largely been wiped off the face of the earth, are an uncommon threat. Alaska is one of the few remaining places on earth where large predators roam in significant numbers, and where there is the possibility that you can be attacked, mauled, killed, and perhaps even eaten by a large carnivorous animal. To the modern human, this type of danger has a dark appeal that is much more romantic and glamorous than the drunken imbecile weaving across the yellow line and sending you headfirst through the windshield of the family sedan. It is much more exciting to strap on the weapons than to strap on the seat belt.

§

We have just arrived at Denali National Park, and it will be my first real taste of the Alaskan wilderness. The trusty bear gun, transported thousands of miles by rail, air, and highway, will remain in the trunk of the rental car. No guns are allowed in Denali National Park. If a bear decides to kick your ass, it's all part of the primordial experience.

With our backpacks leaning against our legs, we stand and watch as the bus pulls away, dwindling to a shimmering dot at the base of a feathery plume of dust. The shuttle buses, which run the two-hundred-mile round trip to Wonder Lake and back, are our only connection with civilization.

What a strange National Park this is: no automobiles! You actually have to get out and walk! What a concept... My one and only trip to Utah's Arches National Park mostly involved dodging crowds, although I did finally manage to find some peace when I stumbled across a rough jeep track that the Park Service had somehow forgotten to pave. Driving to the end of that road got me free of 99 percent of the Park's visitors; walking a quarter of a mile into the trackless desert put the human race behind me. If all our National Parks could somehow reject the automobile and the motorized tourist, we could visit them and depart with our sanity intact.

Before lifting the packs to our shoulders, we stand for a long moment and listen to the silence. The tiaga (a Russian name which translates to "Land of Little Sticks," an ecological zone which includes a sparse forest of short, spindly trees) stretches up before us, rising to the foot of the smooth hills to the north, all wreathed in the smoke of a brush fire burning somewhere unseen. The smoke limits our view and distills the weak sunlight into a diffuse vaporous glow. Even so, the immensity of the landscape is intimidating.

As we walk across the tiaga, I can't help but contrast everything I see, and experience, with my home state of Arizona. This is like visiting another planet. For one thing, there is water everywhere. Every few hundred feet we cross a rivulet or small creek. Between the creeks, in every swale and dip in the land, are bogs of standing water. As we splash through these tiny swamps, our feet disturb thousands of mosquitoes, which rise to swarm around our legs. The air is humid, and soon a cool mist is falling. After half a mile, I stop and sheepishly empty the three canteens I have stowed in my pack.

"I can't believe you brought all that water! This is Alaska, dude," Lisa laughs.

"I can't help it. Going on a hike without water is alien. I'm a desert rat." I take a long, slow look around. Water gleams on hillsides and drips from the dark stunted trees. The high peaks of the Alaska Range, invisible as usual behind a bank of clouds, are forever cloaked in snow and ice. Water.

"This is wet like Hawaii, only you can tell from a glance it gets incredibly cold here in winter." Looking at the dwarf trees makes me shudder. Even though this is high summer, winter rules here. The diminutive trees are evidence of its power.

"Oh, yes," she replies, "very cold. Especially here in the interior. On the coast, the ocean keeps things from just freezing solid, at least. Here, the arctic air comes down, and stays. Everything freezes."

In the winter the ground we walk on will be frozen hard to a depth of several feet. In summer, the relative warmth has thawed the surface of the permafrost, and the ground shimmies and shakes like gelatin as we walk.

After we make our first camp, I climb up on a little knoll with topography map and compass, intending to fix our position. Twenty minutes later I have managed only to become completely confused. I know our general position, but fixing a position requires triangulation, which in itself requires recognizable landmarks. I can recognize nothing.

Frustrated, I go back down to our small camp and retrieve my binoculars. In the misty distance, through the powerful binoculars, I make out a smooth, dome-shaped hill just to the north of a bend in what looks like a small watercourse, and suddenly the landscape becomes clear. I am not using my usual seven-and-a-half-minute series map (1:24,000 scale, or 1 inch = 2,000 feet). It's a 1:63,360 series map (one inch on the map equals 63,360 inches, or one mile).

Alaska is big. The line of smooth, massive hills to the south isn't two miles away, but five. The river bottom to the north isn't a small canyon bottom; it's a full mile across! Big state, big country: big maps.

Two days later, the vastness of Alaska still has me in awe. We are moving downstream in the river basin of the East Fork of the Toklat River, hoping to cross. The going is excruciatingly slow. The south bank of the river, and the country to the south for miles is covered with brush so abundant and close that it makes walking a time-consuming and exhausting chore. To the north the country opens up invitingly, tempting us to explore its mysteries.

"I never pictured tundra being covered with brush like this. No wonder the locals use the river bottoms to get around," I mention to Lisa as we prepare to cross the river.

We take off our camouflage army surplus trousers and put on nylon rain pants. Our boots are tied to the outside of our packs and replaced with old high-top sneakers, or "river shoes." Before stepping into the water, we release the waist belts and chest straps on our packs and loosen the shoulder harnesses. If we go down in the swift water, we've got to be able to shed the packs in a hurry.

"Yowww!" Lisa yells as the water seeps through her river shoes and two pair of socks. "This isn't water, it's ice!"

The current is swift, and the water temperature feels like thirty-three degrees. Maybe thirty-two point five degrees. My legs up to the knee are numb within seconds. Fifty feet into the calf-deep river, our necks and shoulders begin to ache as the river just sucks the heat out of our bodies. A hundred feet out and the dull ache begins to become real pain. We cannot see the bottom of the river, even though it is perhaps just eighteen inches deep here. The water is full of sediment, and we are in constant danger of losing our balance with the heavy packs shifting around on our backs as we struggle for footing on slippery, unseen rocks.

We retreat.

The sixty-degree air feels like a sauna as we roam up and down the river, looking for a better place to cross and to find me a walking stick. I finally find a five-foot length of gray driftwood that looks like it might do the job.

"This river is a lot bigger than it looks on the map," I complain. "It must be a hundred yards across! It's full of dirt, or sediment, or something. It's tougher when I can't see where I'm putting my feet. This stick will come in handy."

"The cloudiness is from glacier flour, or something like that," Lisa informs me as she rubs circulation back into her strong, muscular legs. "The glaciers grind the rocks into a powder as fine as flour. That's why there are no fish in the rivers here. I'm not sure fish could live in water this cold anyway."

"Where are the glaciers?" I ask, jumping up and down to try and get some feeling back into my feet.

"Up there." She points westward toward the tremendous peaks of the Alaska Range, the lower slopes of which are visible below the almost constant cloud cover. "This is glacial runoff."

"This river is?" I can't believe it. "This whole river? This is a big river!"

"The glacier up there melts in the summer and makes the river run." She pauses. "Evidently someone wasn't watching the film on river crossings." She smiles at me. "I was, and that's why we know to put on our rain pants for insulation."

"No, I was looking at the photos of bears. Besides, I've crossed a few gnarly rivers before. This is just like any other river, except it's bigger and wider, and colder. A lot colder. That explains why the water is so cold. Just up there, it's ice!"

We put on our packs and enter the river again. The strategy is that Lisa stays on the downstream side of me, pushing against my back with one hand to help keep us upright. My body breaks the current for her; she holds onto my arm with her other hand for support as I probe ahead with the stick. Cautiously, one careful step at a time, we move across the braided river, emerging from the current several times to cross small sand bars, and then dropping back into the next braid of fast water. The stick helps a great deal, and the going is fairly smooth until we reach what appears to be the main channel, where the water climbs halfway up my thighs. The current is suddenly very powerful, threatening to knock us off our feet, and I have to lean on the stick time and again to keep us both upright.

The piece of driftwood flexes ominously every time I put weight on it. The water is paralyzingly cold. By the time we have gone forty yards into the river, my neck, shoulders and lower back are throbbing painfully. The water is roiling around Lisa's waist, tugging hard at the bottom of her pack. She has to reach around with one arm and hoist it up while hanging on to me with the other.

Suddenly, I lose my balance and nearly fall. Lisa's firm pull on my arm and a desperate stab with the stick barely keep me upright.

"Don't go down!" Lisa says. "If you go down, we both go down. I would not want to go under in this!"

Although I have tried to conceal it with a stream of reassurances and manly curses, I am very worried as well, and have been for the past several minutes. This is suddenly not just a bit of adventure where there is a risk of some small injury. Going down in this water could mean death. The river would just roll you and roll you and, being in shock from submersion in the icy water and wearing a heavy backpack, you might never get up again.

"This is dangerous," I finally tell her. "Fine. I'll be the big wimp. Let's turn around now, while we can."

"Yes!"

Just as carefully as we went in, we go back across the silver shimmering water. It would be beautiful if my head didn't hurt and I could feel my feet. Like a big hungry icy beast, the river pulls the strength from our legs. Concentrating on my balance, I have to take the time to deliberately set my legs in a solid stance at each step. From time to time I am supporting half of Lisa's weight plus the weight of her pack on my left arm. Step by step, encouraging each other with a word or a hard, quick squeeze of a hand on a forearm, we shuffle into safer water. By the time we finally stand on solid ground, Lisa's hands on my forearm are a vivid blue.

Too cold to enjoy the blessed feeling of relief, we clomp downstream on our frozen shaky legs looking for a campsite. Finding a decent place on the south side of the canyon, we discuss the failed crossing as we set up camp.

"Well, now I know why they recommend that only experienced hikers go into the backcountry here. Still, I wish we could have made it across." Lisa the adventurer is disappointed.

"I'm actually very proud of us," I reply. "We were experienced enough to know when to turn back."

The great thing about summer hiking in Alaska is that at the end of the day, you have another full day of sunlight ahead of you! After the tent has been set up, I look at my watch and am astonished to find that it is 8:00 P.M. The sun isn't even close to going down; it seems like mid-afternoon.

We are both ravenously hungry. What we would like to do is rip open our packs and prepare a meal on the spot, but one doesn't do that in bear country. Not if one is prudent, that is.

The authorities at Denali National Park are quite concerned about the unique (by modern standards) situation of two dominant animals sharing the same space, however vast. Through years of studying the great bears that roam the Park, they have come upon a strategy that seems to be working. The first part of the strategy is to limit the number of people in an area of the Park at a given time. The second is to teach good bear etiquette to those who are hiking into remote areas. And the third is to never give bears the notion that people are a source of food. Grizzly bears rarely attack humans, and far more rarely still attack humans and eat them. Bears are, however, attracted to the food that people have with them, and if hungry, may take it. This can very quickly give bears the idea that people are a good source of food, possibly developing into a habitual pattern of camp raiding, which can obviously have tragic consequences. Using a technique that was pioneered in other National Parks such as Yosemite, the clever people who run the Park have devised an ingenious system to separate people from bears by separating people from their food. When backpacking into the backcountry, hikers are given a Bear Resistant Food Container (BRFC), a stout plastic cylinder with a tightly fitting cap. All food, garbage, toothpaste, gum and anything else that might be attractive to a bear goes into the container. After making camp, the hikers are instructed to take their containers of food a minimum of a hundred yards from their camp and leave it. When preparing a meal, the hikers then take their BRFCs and go to another spot a minimum of a hundred yards from the food storage area and equally remote from camp.

For Lisa and me, preparing a meal involves:

1. Walking a hundred yards to the containers.
2. Taking the containers and walking another hundred yards to a suitable place for eating.
3. Preparing, eating, and cleaning up after the meal, being careful not to spill food on the ground, or, more importantly, on our clothes.
4. Walking back to the food storage site, where the BRFCs are again deposited.
5. Walking back to camp and very carefully washing our hands in the river.

This system is an excellent way to lose weight. If you want a snack, or even a stick of gum, it involves a quarter-mile walk. It's a pain in the neck, but it beats having a bear in your tent.

As we eat, I ask Lisa to tell me a bear story. It's a story she's told me many times and one I never tire of hearing. It's about a man she worked with at the fisheries in Seward, Alaska.

One day after work, this man and some friends had gone fishing. After a few hours of fishing, they returned to their tent and made ready for sleep. Having been successful fishermen, and not wanting one of the numerous bears in the area to get their fish, they brilliantly decided to safeguard their catch by putting the fish in their tent when they went to bed that night! In the middle of the night when they were sound asleep, a prowling black bear came into their camp. Our hero had his head pressed against the thin fabric side of the tent, and the curious bear gave the round object a sound swat. Thus rudely awakened, the men poured out of the tent, and the startled bear promptly chased them down a cliff (the fate of the fish is unknown, but we can assume, and hope, that the bear ate them all). This same fool was also treed that very summer by an enormous cow moose that he decided to frighten. Some animals don't frighten well; if annoyed they have a disquieting inclination to run straight at you. How unsatisfying for him. (The moose probably thought she had a good day; although our hero escaped her hooves, the mosquitoes liked to chewed his ass off).

Lisa and I spend the next day, the next long and shining and most glorious day, roaming up and down the river exploring side canyons. Early in the morning we are working our way downstream trying to keep our feet dry and I am grousing to Lisa about how, on its first use, I had promptly clogged our expensive water filter with the ubiquitous glacial flour. Cleaning the filter had included hiking to the food container, collecting my toothbrush, walking back to the river, and laboriously scrubbing the filter canister. After several cycles of cleaning and rinsing with clean water that I found in a quiet pool, the filter was operative again. And then I made another two-hundred-yard round trip to the BRFCs and back, only to discover I had forgotten to brush my teeth!

Looking up, we spot a beautiful red fox coming upriver against the south bank. As it trots nearer, we can see that it has a small ground squirrel in its mouth. When it is about thirty yards away, it abruptly drops its prey and darts up and over the bank. We are both convinced we frightened it away, but in a few seconds, it returns with another rodent in its mouth. Amazed, we watch as the small fox manages, with some difficulty, to ram both critters into its mouth. Head up, cheeks bulging and both animals hanging out of its mouth, it trots right by us with scarcely a glance.

As the day passes, we often climb the steep riverbank just to sit and drink of the incredible landscape. Hooded and cloaked in dark cloud, the great mountains of the Alaska Range cannot be seen, but their presence is felt like a massive crouching animal barely seen in a dark forest. Occasionally, the clouds part just enough to give a tantalizing, shadowy glimpse of the lower slopes, and the mind tries to fill in the rest. Is that impossibly high white glimmer a cloud, or is it the shoulder of a mountain too large to be true?

The wide braided river runs down from the mountains, a lustrous silver and gray, lying like a long necklace against the brawny chest of the land. When we have filled ourselves to the brim and it is time to move on, we walk with caution, eyes up and all senses on full.

Bear country! It puts a snap in the step and a sparkle in the eye, that eye watching the skyline and the backtrail. This is fundamental terrain: big, open healthy unscarred wilderness. Wolves and giant bears roam these hills as they have for thousands of years. Too cold for cattle and corn, sufficiently remote to allow its protection before it could be subdivided or strip-mined, the full array of primal opportunities is still here in Denali. This part of Alaska is as wild as everything used to be.

§

Late in that same day we happen upon a flat-bottomed side canyon that is an Eden of wildflowers. Lisa is in a state of bliss; the flower girl has found a natural garden. Leaping the tiny creek, we wander upstream through the canyon bottom, enjoying the brilliant display of yellows, greens, purples and flaming reds, all in startling contrast to the misty expanse of blue green tundra. Wildflower manual in hand, we identify the red Mountain Sorrel (Oxyria digyna), the delicate white bells of Labrador Tea, beautiful and deadly (if eaten) purple Monkshood (Aconitum delphinifolium) and purple-flowered Eskimo Potato (root edible when cooked). Inspecting a particularly spectacular stand of fireweed (Epilobium angustifolium), we surprise a porcupine, which makes its slow and deliberate way up what is evidently the only tree within twenty miles ("Cottonwood tree," Lisa advises.).

And everywhere, of course, there are the mosquitoes. With all the talk about the dangers of bears and the awesome charge of the cow moose, the most terrible creature in Alaska is the mosquito. As soon as we leave the breezy river canyon Lisa is forced to wear a head net; the little monsters find her irresistible. We wear three layers of clothes, and not for the weather, which is fine and warm, but as armor to turn away thousands of insatiate probing lances. Some of the mosquitoes are astoundingly large, twice the size of any mosquito I've ever seen. We start calling these the "Grizzly Skeeters." Insect repellent helps a great deal; without it head nets and gloves would be a constant necessity, but it in no way provides total relief. They bite us through our hair, through our clothes, and through the tent, if we are unwary enough to rest an arm or cheek against the nylon wall. The high-pitched whine around the ears is maddening because it never really stops, and it sounds so very hungry.

Many years ago, I once engaged in conversation with an officer of the Bureau of Land Management. It was a chance meeting on a lonely road under pink sandstone cliffs in southern Utah. Our pickup trucks parked nose to nose, we leaned on the hood of his battered old Dodge, drinking warm beer and talking about the canyon country.

The beer was a gift from God. Looking for a map to start our discussion, I had poked around under the seat, rummaging through leaves and oil cans, feathers and spent shotgun shells, and there was the magic six-pack. The beer had been aged to perfection through months of being banged and bashed around on rough old dirt roads and being heated again and again to temperatures well in excess of the recommended maximum storage limit. It was a wonder that the brown bottles hadn't burst, but then again, God had surely intervened. And only God knows how long the beer had been under the seat of my truck. I was sure it wasn't there when I bought the truck five years earlier, but everything after that was kind of hazy.

My new friend, whom I never saw again after that day but think of often, broke off his study of the map to remark, "Damn, that's good beer! I wonder what it is?"

We were never to know the brewery. The labels had been almost rubbed off.

The breeze died down, and the gnats swarmed us like little black bits of burning coal. Inspired, no doubt, by the magic beer, my friend made a most profound statement: "Dammit! I hate it when they bite me between my fingers. You know, right there?" pointing to the tender web between the fingers of his off-beer hand. "In fact, I just hate gnats. Once had one get up my nose, and I'm not kidding, I had the can of WD-40 right there, ready to shoot it up my nose at the little bastard, it was all that I had, you know? But he quit kickin' around, must have suffocated. I love the outdoors, but do you ever wonder what the little bastards do when they're not biting you? I mean, what the hell do they do?"

He was a nice man, that government officer, but I sometimes wish I'd never met him. His comment inspired a vision; a terrible image that had never come to me before that moment; thousands, millions, billions of bugs clinging to the dusty bark of cedar, the silver sage, and the needles of the pinyon pine. Glassy, multifaceted eyes focused on forever, with not a thought in their empty little bug heads; mindless creatures waiting for me to walk by so they can fly out and bite the corners of my eyelids. And that vision has remained with me for years, arising from time to time, and I can tell you that it is a chilling specter of darkness. Dwell too long on the thought of four billion stupid mosquitoes hanging upside down on the damp leaves of the dwarf birch, driven by their ancient instincts to drive a fiery needle into your flesh, and you will go stark raving mad: Bring on the grizzlies! I'll eat 'em raw and string their claws for a necklace! Just keep the bugs off me! Aaaaaaaaaagh!

While we sit on the lush grass at a bend in the tiny creek, Lisa produces a pamphlet from her daypack. Lisa loves pamphlets. Any piece of paper with a scrap of information on it will be swept up and saved for later consumption, rather like a squirrel collecting nuts for the winter. Our house is filled to overflowing with these little kernels. Peering through her mosquito netting, which makes her look like The Fly in hiking boots, she reads: "Brown bears often roam many miles in search of food. In the Alaska interior, river bottoms are often heavily used travel routes by bears. These bear highways..."

"Bear highways! No... We're camped right on a big river!"

"That's what it says..."

"Great."

§

We wake up in the morning to the chuckling of ptarmigan. There is a whole group of these birds (Or is it a covey of ptarmigan? A babble of ptarmigan? A stupid of ptarmigan?) just outside the tent. Still in our sleeping bags, we crawl forward and peek through the mosquito netting. They are pretty birds, basically a large brown and white grouse. When I crawl out of the tent and stretch, the birds amble away, not particularly worried by my sudden appearance. I take a step forward and there, not six feet from the tent, is a colossal bear track. My size ten boot is maybe a bit longer, but not nearly as wide, and... it wasn't here when we went to bed!

"That's it!" I say. "We're moving camp! The Arizona fool and the California girl, here we are, in the damned wilderness, babes in the woods!"

Lisa tries to calm me down ("It was only one bear and it didn't bother us," etc. etc.), but it's no use. I'm plugged in. After a few minutes of this nonstop paranoid raving ("They won't let us bring guns in here. No wonder the bears are so fat and sleek! All these little unarmed bear snacks running around!"), she's over the edge, too. Soon we're both running around and yelling. Disturbing the pristine peace of the morning. Scaring the wildlife. Even the ptarmigans shuffle off to quieter environs, chucking disapprovingly down in their feathery throats.

Looking for a campsite off the river bottom, we scramble up the steep bank. I look up, and there is Mr. (or Mrs.) Griz, eating berries.

§

Lisa and I spend several more physical and joyous days along the river before packing out. After reaching the point where we had been dropped off along the long dirt road that runs through the Park, we decide to ride the bus all the way to Wonder Lake, the end of the line.

At the next stop, four young men climb aboard carrying backpacks and looking as though they have spent a few rough days in the bush. Someone asks them how their trip went, and one of the young men wordlessly pulls the remnants of a tattered tent from his pack.

The bus explodes into an uproar. The long, nasty rips in the fabric can only have been made by one creature.

"We decided to come out a day early after a grizzly got our tent!" he laughs. It is a somewhat hollow and distant laugh, made by one who seems happy, maybe for the first time in his life, to be on a bus.

"Trashed it, man," another of the hikers says, "like a werewolf got it. We packed up and took off."

Fifty people ask the same question at the same time. The offending bear was nowhere to be seen as they hiked out, they answer.

After a few minutes the bus quiets again, and the bus driver, in a tone of complete authority, says, "Sounds like the Tent Bear is back. Or, more likely another has taken his place. After what they did to him it's not likely that the Tent Bear would ever destroy another tent, or even want to come close to a tent."

He has our complete attention. If it weren't for the groaning of the old bus and the roaring of the diesel engine, you could hear a moose snort a mile away.

Relishing the drama, the big blond driver waits until somebody asks, as he surely knows someone will, "What is the Tent Bear?"

And fifty people lean forward on the hard seats to hear the tale of the Tent Bear:

A few years back, in this same area that these men were hiking in (And Lisa and I; these four hikers with the mauled tent were on the north bank of the river we had tried to cross a few days earlier), a big male grizzly got into a very bad habit. Perhaps it was just something about the shape of a modern backpacking tent, or the smell of humans, or maybe it was just the sensual pleasure of feeling a tent go to pieces under his eight hundred pounds. Whatever the reason, the Tent Bear, as he soon came to be called, liked to destroy tents. Backpackers would return to their camps to find their shelter ripped to shreds and mashed as flat as a dinner plate. No food was ever taken and no one was hurt, at least not at first.

The authorities were concerned about this strange and aggressive behavior, although certainly not nearly as impressed as the bedraggled, frightened, and mosquito-bitten hikers that staggered into the Visitor Center to tell their tales. The level of concern jumped a few notches when, after a brief period of summer calm, the Tent Bear struck again. There was nothing exceptional about this particular tent, aside from the fact that it happened to be in use at the time. As the bear was not interested in attacking people, just their offending tent, the occupants of the tent fortunately escaped with only bruises, abrasions, and, one can assume, soiled underwear. At this point the authorities were faced with some tough choices. Bears destroying property go into one category. Bears injuring people go into another, and this second category almost always assures a death sentence for the bear. In this case, however, because the people inside the tent were injured by accident, and were not the targets of this somewhat neurotic bruin, the authorities decided to pull a sting operation on the Tent Bear.

A modern backpacking tent, a sleek synthetic dome with fiberglass exoskeleton, was erected on the bank of the river where many of the Tent Bear's crimes had been committed. Inside the tent went three men, equipped with radios and armed with rifles. A fourth man was positioned on a ridge overlooking the tent. This man was also equipped with radio and high-powered rifle. Two of the men in the tent loaded their rifles with rubber bullets. In case the sting operation went sour, however, the ridge watcher and the third man in the tent had their weapons loaded with big-game hunting ammunition.

The setup proved to be too much of a lure for the Tent Bear. The men inside the tent, nervously waiting for a call on the radio and fervently hoping that their man on the hill stayed alert, suddenly got an excited message that a bear had been sighted and was coming in on the run. Proving themselves to be steady of nerve, the three men waited until the charging bear was within fifty yards of the tent. Leaping out, two rifles pummeled the bear with a fusillade of rubber bullets. The bear swapped ends in an instant and streaked out of sight. The reign of terror of the Tent Bear had ended.

§

The day is very clear, and when the bus stops in a low pass well out onto the high, grassy tundra, we are treated to our first view of Denali, also called Mount McKinley. Denali ("The Great One") rises from two thousand feet at its glacier-carved base to over twenty thousand feet. It is a muscular young mountain, jagged and well defined, and it goes up right now. What you see is three and a half vertical miles of rock and ice, eighteen thousand feet of raw power. The tundra is green and warm, immersed in the brief bliss of subarctic summer, but it is cold up on that mountain. It is always winter up there. Great tongues of ice slide slowly down its colossal flanks. High velocity winds screaming over the summit pick up snow and ice and hurl it into a horizontal sheet that stretches a mile or more from the peak. The lesser giants of the Alaska Range surround the mountain. The immense dome of Denali humbles these sharp, ragged peaks of ten, thirteen, and even seventeen thousand feet, imposing and regal in their icy blue vertical silence.

Lisa and I stand quietly, looking at the mountain.

"It was worth the long trip," I finally say, "just to see the bear." Pointing up at Denali, I add, "And that. And I don't know which is more impressive, the bear or the mountain."

"It's all the same," she says. "Alaska!"

BIBLIOGRAPHY

Green, Martha Hodgkins. "*Continental Divide*." The Nature Conservancy. January/February 2000.
Peacock, Doug. *The Grizzly Years: In Search of the American Wilderness.* Henry Holt and Company, Inc., New York, 1990.
Pratt, Verna E. and Frank G. *Wildflowers of Denali National Park.* Alaskakrafts, Anchorage, Alaska, 1993.
Treadwell, Timothy. *Among Grizzlies: Living with Wild Bears in Alaska.* Ballantine Books, New York, 1997.
Alaska Department of Fish and Game, informational pamphlet titled "The Bears and You."
Denali National Park, backcountry hiker training videos.

JEWELS IN THE DESERT

I. The Great Canoe Trip

As the author of this work I'll take the writer's prerogative and blame Tom. It was Tom who had the idea for the canoe trip. I also recall that it was Tom's idea to use his father's handsome and venerable canoe for the voyage. After all these years I still feel guilty about what we did to that poor canoe. I mean, what Tom did.

Toms' father, Tom Sr. and his wife Consuelo had retired to one of the ultimate places to live, that is if you love desert hills, desert towns, and desert ways. Their house overlooks the San Francisco River, which may be one of the last relatively unimproved desert rivers in the country. Unimproved, that is, in that it just runs from its headwaters down into the Gila River, free of dams, canals, ditches and federal employees.

One morning, as Tom Junior and I walked off one of Consuelo's marvelous and prodigious breakfasts, we happened to open the garage door and for the first time I lay eyes on his father's canoe. I didn't think much of it at the moment, being fixated on the glow of a mighty breakfast, the pretty morning, and the shining little river below. Besides, I've never been much of a canoe man. Had some poor experiences in canoes, actually. Fallen out of a few. Unstable craft. Very narrow.

As a boy, I watched in horror as my father, who was as experienced in small boats as any ex-navy outdoorsman could be, dropped a magnificent Winchester Model 12 shotgun into a lake. From a canoe. Due to unfortunate circumstances my father and I were forced to hunt ducks that season from a borrowed canoe. Forced by which I mean it was duck season and Dad's boat had a bad leak. One who does not hunt ducks might scoff at the term "forced." After all, this type of person might argue, no one was making us hunt ducks under pain of death or anything. This type of person is impossible to argue with, but duck hunters understand. If the choice was to hunt ducks from a canoe or not hunt ducks, then one is of course forced to borrow the canoe.

Well then, to my father. For all of his hunting life he had customarily set his shotgun in the boat to his right hand, leaning it into the corner made by the seat and the side of the boat. Most boats are wide. Canoes are not wide. One icy winter morning, on a small lake in eastern Arizona, two mallards had lifted from the reeds just offshore. Dad brought his gun up, thought better of the long, difficult shot, and then casually set his shotgun in the lake.

Dad and I both loved that gun. We loved that ancient shotgun in the way that only those who have held shotguns on icy winter mornings can, with the ducks inbound and the old Labrador retriever warm against your leg. In a twisted sort of way, I always sort of blamed the canoe for the loss of Dad's shotgun.

I sometimes wonder if the canoe trip was a subconscious attempt at revenge.

§

Tom made some casual remark about the canoe as we leaned on the short chain link fence beside the garage, chewing on toothpicks and watching the river. Our eyes strayed from the river to the garage and back to the river, shimmering in the early morning sun.

"Dad loves that canoe," Tom said, "he's had it for years."

"Not much of a canoe man, myself," I replied. "Prefer rafts or inner tubes. More exciting."

"Canoes are a blast! Have you ever been in a canoe?"

"Of course I have! Many times. Dad and I used to hunt ducks from a canoe."

"Spent many a day in a canoe," Tom countered.

"Yeah, back east. The rivers are a tad different out here."

The gauntlet thrown. Tom was raised in the Southwest, but as I well know he is extremely sensitive about being incarcerated for several years in Indiana. It is a nice place, Indiana, but it is not the Wild West. There are no towering crags, and no savage river canyons, in Indiana.

Tom shuffled his feet and took a better bite on his toothpick. Then he lowered his head (just like a damned badger) and looked at me out of the corner of his eye. I decided to cool him down before he could retaliate.

"Be a helluva good time..."

After a pause, he replied, "Better believe it. The 'Frisco empties into the Gila River, and from there you can ride the Gila all the way to Safford. I've always wanted to do that."

"All the way to Safford? Guess I never really thought about it. You're right!"

"Take a few days. Three or four, if you wanted to do it right and see the country."

"Through all that amazing wilderness."

"Stick," he drawled in his best "lemme-tell-ya-son" voice, "you would not believe the country that river runs through. Incredible. Wild!"

We stared at the river.

"Dad loves that canoe," Tom said again.

We went inside the garage and looked at the canoe. It lay upside down on a rack. Its aluminum hull was amazingly smooth and sound for a vessel its age. The seats and wooden paddles were well kept, all in all an attractive craft. It had an expensive look about it.

"He hardly ever uses it," Tom said. "Hell, it just sits in here. We could put it to good use."

"Ask him if you can borrow it."

"I'm afraid of putting a ding in it. It's an aluminum canoe, kind of a lake canoe, you know?"

"We wouldn't necessarily hurt it. I mean, it might pick up a scrape or two."

"The Gila River has a lot of rocks..."

"What about just starting from here? Going down from here, all the way to Safford?"

We went outside and stared at the river. From where we stood with the morning sun just so, the 'Frisco seemed deep and wide.

§

Leery of asking Tom, Sr. if we could borrow his prized canoe, we just decided to borrow it for a few days and return it after the trip with no one the wiser. We would take excellent care of the canoe and paddles, going slow and easy on the rivers and portaging if things got rough.

This trip was to be the preliminary run, an exploratory journey, a test of the craft and our mettle before the Big One. The Big One was the long float down the Gila, or maybe the Verde, taking as long as we damned well pleased, fishing for big catfish and hiking our legs off. My fantasy was to live like an educated savage, sort of a combination of Huck Finn and the Desert Rat.

We were always planning the Big Ones, the gnarly trips into the last wild places of the American West. Sometimes we got past the planning stages and went, and at times we just had to dream. Planning a voyage on a navigable length of wild river suitable for raft or canoe in the United States of America takes a bit of searching. If you are considering poor dry Arizona (where rivers are scarce and land developers are many) for your adventure, you will have to study a map of the state with determination. And simply studying a map is often not enough; long stretches of the lower half of the unfortunate Gila River, for example, are often still drawn as if water actually flowed there. It once did. Before that horde of crazed beavers (our U.S. Bureau of Reclamation) attacked nearly every river in the American West in a colossal attempt to turn the deserts into something resembling Iowa, the Gila was a living river for its entire length. West of Florence, Arizona, the Gila River has been killed to irrigate crops and its bed is completely dry except in times of extreme precipitation. The lush riparian oasis, fed by a river teeming with trout and other fish, has long since died and blown away.

§

A few weeks later came the long Memorial Day weekend, and with Tom's father safely in Tucson, the canoe was loaded into Tom's truck and driven down to the river. We had originally planned on carrying the canoe and supplies down to the river from the house, but we soon discovered that the canoe weighed as much as a small cow. We felt only slightly like thieves. The canoe was beached, the gear was loaded, and then the now-legendary discussion of who would sit in the stern began. Over the years I have been reminded many times of this conversation, and it has been forever burned into my brain:

Tom: "The best oarsman should always sit in the stern of the canoe."

Me: "Why?"

Tom: "Because the one in the back steers, Jeff. Steering is very important."

Me: "I know that."

Tom: "I will sit in the back. I have some experience in a canoe."

Me: "On what, a lake or something? We will be running rapids! I have whitewater experience. On the Upper Salt. Nasty river. I should do the steering." (I neglected to tell Tom that despite my frequent trips down some truly rough water, my experience had mainly consisted of clinging upside down to a half-inflated raft or inner tube while getting the stuffing knocked out of me by giant rocks.).

Tom felt obligated to concede. His Midwest canoe experience just couldn't go up against my rugged, manly whitewater seasoning on wild western rivers.

Anticipating a peaceful float, we are, of course, wearing no life jackets or helmets (but remembered to splash on a bit of sunscreen, got to watch those fierce rays) and have no spare paddles. There is nothing in the human catalog that can create adventure, and even danger, quite like being thoroughly unprepared for the task at hand.

We shove off, and for about five minutes things go well. For the first few hundred yards the little river runs straight and true and we shoot down the center like an arrow. After a couple of comfortable twists and another long straight stretch, the stream sweeps right, then makes an abrupt left turn. I'm not talking about a gentle bend here, but a sharp wicked curve of a hundred and twenty degrees, with the river narrowing to about fifteen feet wide and going like crazy over big rocks. At the very apex of the curve the river shoots under a deep overhang, which clears the bow of the canoe by about a foot.

"Okay, Stick," Tom says calmly, "start making your turn."

"Sure thing!" the manly river man replies, and then sets to flailing ineffectively about with the paddle, which, in fact, is the only thing I know how to do in a canoe.

"Turn now, Jeff. Turn now!"

"Okay. No problem..."

"Turn! Turn! Aaaaaagh!" adds Tom, diving for the bottom of the canoe.

The canoe shoots straight into the overhang. I have time to flop back against the stern and tilt my head back almost in time to miss the craggy overhang. My head is simultaneously slammed against the aluminum stern of the canoe as very rough rocks take the skin off my nose and scalp. The canoe flexes and rebounds into the current with a shuddering crash. Tom is nearly thrown out of the canoe by the impact, and has to drop his paddle into the canoe and scramble back in.

"Take her about!" Tom yells as we enter the stream sideways. The stern catches a rock and comes about smartly.

"Got her!" I yell triumphantly.

Tom handles the canoe expertly from the bow, taking her through some minor ripples as I whip the surface of the river into a fine froth with my paddle.

"That was close! If that overhang would have been any lower it would have ripped my nose off!"

"You need to be a little quicker on those turns, Jeff," Tom replies, "and here we go! Turn! Dammit!"

Wham! The canoe hits the rock wall at the back of the next turn broadside, throwing our gear and both of us hard to the right. Thrown nearly out of the canoe, I reach out my right arm to steady myself against the cliff, but the canoe has rebounded a few feet out into the current. Unbalanced, I go overboard but manage to stay half in and half out of the canoe by clamping my legs tightly around its narrow, curved side. This has the effect of tilting the canoe alarmingly to starboard. Tom compensates by throwing his weight hard to the left, cursing artistically and paddling like a demon. Facing rearward as we enter a rapid, I can do nothing but hang on. My paddle is in my right hand and is inside the canoe, wedged partially under the stern seat. It makes for a good handle, giving me some purchase on the accursed canoe, which bucks like a wild horse in the rough water. The canoe rises again and again to smash my groin as I doggedly hang on, determined not to fall off (the mighty river man fell out of the canoe on the second turn!) and trying not to shriek in agony. The canoe finally bumps into a half-submerged tree, rocking us to the left, and I slide limply into the bottom of the canoe.

"I could use some help, Stick," Tom calls from the bow. His paddle flashes as the canoe bangs off submerged rocks and begins to pick up speed. Tom skillfully maneuvers the canoe through the rapids, and the river settles down.

I manage to climb back onto the seat. "Oh God, I'm neutered," I mumble to myself.

We paddle on in silence. The little river twists and turns like a big snake as it cuts through the desert, digging deeper and deeper toward the Gila River. The canoe repeatedly hangs up on submerged rocks and snags, and time and again we have to hop out and walk the canoe through the shallow spots. It occurs to me that even expert canoeists would not consider trying to take a heavy canoe down a swift little stream like this one.

"I thought you said you knew how to steer a canoe!"

"Well, I do, but this river turns a lot, and..."

"I'm taking the stern. If not, we'll be lucky to make it out of here. End of story. Period."

As the morning wears on, it becomes more and more obvious that the San Francisco "River" is in fact a smallish creek that drains rather steeply, resembles a moist rock garden for considerable distances, and would be considered non-navigable by anyone with a lick of sense.

We have just switched positions and are paddling around a rare gentle curve in the river. From up ahead a low, powerful rumble can be heard.

"Paddle, Jeff! We've got to get over to the right!" Tom peers over my shoulder. "Hear that rumble? This is a big one coming! You need to help me control this thing this time!"

He's right on both counts. The current picks up speed, turning into a brown lunging torrent that makes a series of tall standing waves over unseen rocks. We enter the rapid nicely and the sharp bow cuts through the waves with a steady pounding rhythm: "Thump! Thump! Thump! Thump!" With each wave the bow rises sharply, lifting me high over Tom's head. Despite the throbbing pain in my groin and head, this is total exhilaration. "Yeaaaoooogh!" I scream in delight. Tom is grinning and yelling as water comes over the bow in big sheets of spray. The waves gradually diminish and the river settles into a relatively calm stretch.

We trade grins and a bit of confidence returns. This is why we came. The river relaxes its grip on us for a spell, and as we work downstream, we fall into the easy rhythm and quiet banter of two old friends doing something they love to do.

Tom and I met in the city, but it was a countless number of shared backcountry adventures that forged a friendship. Common ground was discovered while sitting around campfires and digging old pickup trucks out of mudholes. We both have a passion for the outdoors and similar upbringings in small desert towns. Despite having college educations and the required immersion in modern liberal thinking, we retained an appreciation for the rural values we were taught as children, a healthy loathing of large urban areas, and a tendency to view our lives with a backdrop of landscape and geography rather than human invention.

Without reservation or embarrassment, Tom is what I would straight away call a rugged outdoorsman, someone who not only enjoys the wilderness of the American West but also has the physical makeup and mental skills to match the terrain. Great eyes and steady hands make him an excellent shot with any type of weapon. The best I've ever seen behind the wheel of an off-road vehicle, Tom is a four-season hunter, fisherman and backpacker. He's got a farm-boy type of strength, not particularly quick, but strong as a bull and seemingly impervious to weather conditions.

He probably wouldn't consider himself to be an environmentalist. In his mind there is probably too much baggage to carry with that term, but he is nonetheless emotionally attached to the West, and sensitive to the havoc that has been wrought upon it, especially the environmental damage created by the seemingly endless flood of people into the Southwest.

The river turns hard to the left again, plunging steeply down into another rumbling rapid. The rapid is as yet unseen but I can see spray and mist coming off a big fang of rock squarely in the center of the river.

"Left! We need to go left!" Tom yells, gesturing with his paddle. We lean on our paddles to try and bring the bow around, and I watch in helpless horror as the devil-spawned canoe comes right and goes broadside down the river. Water pours over the side, half-filling the already swamped canoe. The bow catches the big rock and the canoe keeps coming around until we are going down the rapid stern first. I look past Tom, who is spinning around to face the current, to see the river just disappear into a deep black overhang.

As Tom whips around his face blanches into a hard grimace, and then he begins his now-familiar chant: "Turn! Turn it! Turn it! Back paddle, dammit! No, on the other side! Jesus Christ, Jeff! You're in the control seat again. Ah, well, we're doomed..."

This time we both dive into the bottom of the canoe as it arrows into the overhang. The stern smacks the back of the overhang with a sickening jar, and the canoe flexes like a big spring, shooting back out into the sunlight. We leap up and grab our paddles, but the strong current drives the canoe back into the overhang. There is no room to sit up and paddle.

"Push off! Push off! We're gonna go over!"

Hunched over, we press our paddles against the rock wall and fend off. It feels as though we've fallen into a cement mixer. The current bashes the canoe against the top and back of the cave again and again. The snarling river and the hollow pounding of aluminum against rock creates an incredible din inside the tight space. I finally just reach up, grab a projecting rock and pull us along. Tom does likewise from the stern, reaching out for the back of the cave wall and pulling hard. Eyes squinted against the harsh sunlight, we exit the overhang at the downstream end.

"Yes!" I scream. "Go!"

"Whirlpool!" Tom yells as the canoe drops into a big hole and starts to spin. "Paddle hard, Jeff!"

I do. Digging deep into the water, the old paddle snaps off in my hand. I look in horror at Tom, who gives me a look of complete disgust as the river just grabs the little craft and flips her over. The roll takes us out of the hole and into shallow water, and we manage to quickly drag the canoe to shore, cursing, bleeding and bruised.

Collapsing on the rocky shore, I try not to look at the wounded canoe. Tom rights it and begins pulling out soaked gear and laying it on the rocks to dry. His motions are swift and angry. We've been on the river an hour, maybe two, and I've got a nice trickle of blood running down between my eyes and another down the back of my neck. Tom is ready to kill me, I've broken my paddle, the beautiful canoe is badly dented, and I will never have children. I lie on my back and look up between the buff walls of the canyon at the piercing blue sky, trying to understand why I love the outdoors so much. It must be the pain.

"Whitewater man, eh?" Tom gives me a scathing glare as he sets his soggy pack upright.

"I guess I'm a little rusty."

"Have you ever actually paddled a canoe down a river?"

"Well...no. Just rafts and stuff. You know, inner tubes."

"No, I don't know! You have to steer a canoe, Stick! You can't just sit in the stern and scream!"

"You weren't doin' much better, canoe man! Besides, you'd scream too if that damned thing was pounding your vitals to mush!"

"You can't steer a canoe from the bow, man! Hello!"

"Well, I know that now!" I glare back at him, defiant. "We should have worn helmets. My head feels like I've been beaten with a big stick."

"Don't tempt me, man. How in the hell did I ever let you talk me into putting you in the stern? We're in it now, buddy! One paddle, and this is a bad little river! God, I hope the Gila is running full."

"Yeah. This was supposed to be the easy part. The little San Francisco River. Poor canoe...sorry, Tom." I feel totally miserable. "If you want to whack me one, go ahead."

"Forget it. Let's get going. We've got a long way to go."

The Broken Canoe Paddle. (Illustration by Jan York)

We repack the canoe and set off again with me sitting sheepishly in the bow. My job is to warn Tom about rocks and other obstacles. Although still rough and rocky, the river calms a bit. During the calm stretches I examine the canyon walls for likely dam sites. Should we survive, I determine to immediately place a call to the Bureau of Reclamation demanding that this slimy little river be submerged behind a gigantic dam as soon as funding can be made available.

Finally in control of the craft, Tom does an admirable job of river running. His skill with a canoe is immediately obvious. The canoe ceases to run into cliffs, goes through the riffles bow first and misses most of the rocks. Realistically, we should abandon this aluminum bovine and simply walk away while civilization is still close at hand, leaving the canoe to be swept into San Carlos Lake with the next flood. Tom, however, bends to his task and just keeps at it. Two turbulent hours later we reach the point where the clear waters of the San Francisco join the muddy Gila River.

The Gila is a welcome sight. After plowing the San Francisco all morning, it looks like the Mississippi, wide and deep with no rocks or nasty canoe-busting logs to be seen. We beach the canoe and take a break.

"We should have started up at the bridge, and to hell with the 'Frisco." Tom motions upstream toward the unseen bridge, which crosses the highway a few miles upstream. "Running the 'Frisco was a bad idea. Not that there's a damned thing we can do about it now."

"Yeah. Rivers only run one way."

We look sadly at the canoe.

"Your dad's gonna kill us. I'm moving to Montana."

Sipping lukewarm water and munching on soggy tortillas, we are astonished when a group of floating cadavers drift slowly into view. What appear to be dead men on inner tubes spin indolently towards us, with their lifeless parboiled limbs hanging limply into the brown water.

"Poor bastards," Tom chuckles.

"Looks like rigor mortis hasn't set in yet," I agree.

It is near midday in late May and the river canyon is like a furnace. A hot, steady breeze flows upriver and the fat tubes are barely making headway. One of the cadavers lifts its head and mumbles a greeting. The others, five in all, lift a hand in our direction. Five heads flop back against wet rubber as they slowly drift away downstream.

"Well, they're not dead yet."

A few minutes later we paddle downstream through the group. The canoe feels like a great yacht as it parts the waters with a neat bow wave, making the tubes bob and spin in the gentle ripples. One of the cadavers lifts its head again.

"H'lo," it croaks.

"What's up, guys?" Tom replies. "Havin' a good time?"

"Oh yeah, you better believe it!" the others agree. "Havin' a blast, man! You bet!"

"You guys have any food?" I ask.

They laugh. "Yeah, couple cases of beer!"

For all I know they drift the Gila River still, bony fingers paddling feebly, one skeletal hand eternally wrapped around a cold one.

We alternately paddle and drift west, losing elevation imperceptibly into the rugged canyons of the Gila Box country. The silted river is relatively placid here, giving us time to think, to enjoy the magnificence of the river gorge and the wonderful incongruity of flowing water in the desert.

To get an idea of the significance of the Gila River, one needs only to study the rising canyon walls. Its value to the region is magnified by the aridity of the area it drains. Desert, desert everywhere, parched and already burning blond and brown in late spring, well before the summer proper is upon it. The Gila is a very small river that drains a vast arid region of 250,000 square miles. Part of the Colorado River watershed, the Gila begins in the high mountains of western New Mexico and drops into the basin and range country of eastern and central Arizona. Its final, and these days mostly empty run ends in the harsh, blistering western lowlands of Arizona.

Upstream of the diversion dams, the Gila is a slender thread of riparian green where walnut, sycamore, mesquite and cottonwood thrive on the runoff from the always-infrequent rains and the more dependable snowmelt from the river's high mountain headwaters. A long meandering oasis for black bear, coyotes, bobcats, mountain lions, deer, elk, hawks and snakes, riparian corridors are estimated by some to harbor populations at least a hundred times greater than the surrounding terrain.

The riverbed below the dams is in ghastly contrast to this harmonious landscape. Every time I cross the long, low bridge on Interstate 10 south of Phoenix, I look hard for water in the bed of the Gila. In hundreds of crossings I have seen water in the riverbed five times, and in four of those I was rewarded with only a sluggish brown trickle, a pathetic string of water flowing in a channel that once carried the rich fabric of a living river.

It's hard to describe the profound depression that comes over me when crossing that half-mile wide stretch of sand. It is a very deep current of emotion, subtle and difficult to identify or examine. The best way to describe it is that I am willing the river to flow again. Willing trees to grow and fish to swim. Willing quail to call as the rising moon beckons the nighthawks up to swoop and roar, people dancing below on the cool damp sand...

But I open my eyes and there is this long white bone lying in the desert.

The problem with this river, I explain to Tom, is that the Apaches aren't here anymore. If they were still living up there in those hills there wouldn't be any dams downstream, or cattle fouling the water upstream. Practicing open warfare on one's neighbors is certainly an extreme form of environmental protection, but the Apaches and other aggressive and independent tribes along the Gila kept the area largely free from development for centuries. The Apaches arrived in the Southwest not long before the Spaniards began to explore the region, and from that time up until the 1870s very few dared to even pass through the Apachería.

Tom is quick to point out the positive side, at least from our viewpoint, to the end of the Apache Wars: we are free to float down the river without sprouting arrows. One hundred and fifty years ago we would not be journeying down the Gila in this casual and carefree manner, not just the two of us and certainly not in a boat. We would be part of a large group, mounted on horses and armed to the teeth, with our eyes constantly on the backtrail and the skyline. But damn, it would have been incredible! There would still be beaver in the river then, and grizzlies roaming the mountains. The lush forests filling the riverbed had not been trampled and gnawed to the ground by thousands of cattle; the thickets were home for millions of animals and the river teemed with fish. The sky was completely free of photochemical haze and not a dam was to be found along the entire length of the river. At night the wolves would howl, silencing the coyotes...

The demise of the Gila River and the land it drains began in earnest after the Apaches were militarily defeated and forced onto reservations. The majestic timber belt of Arizona's central highlands was clear-cut for miles. John Bourke, an officer with General Crook during the Apache Wars, wrote of his Arizona experiences in his classic "On the Border with Crook," published in 1891. A keen observer and excellent writer, Bourke described the forests of the Mogollon highlands during the 1870s as an "immense pine forest" and the "forest primeval." A scant twenty years later these forests had been "raided by the rapacious forces of commerce" to a degree that obviously alarmed the old campaigner, and that was over a century ago.

Enormous mining operations were soon in progress, and huge herds of cattle were driven onto the grasslands. The end of hostilities also brought a flood of people into the Gila drainage. Some of these people were farmers, and within a few years the Gila had been dammed to irrigate crops. By the mid-1870s the descendants of the ancient Hohokam civilization, the riverine Akimel O'odham, people who had lived along the Gila for centuries and demonstrated an admirable expertise in farming (one year the Akimel O'odham sold the U.S. Army five million pounds of surplus wheat!) suddenly did not have enough water to irrigate their crops. The end of that decade saw the death of the lower river; the dams near Florence had taken almost the entire flow.

If the trend continues, all running water in Arizona will be siphoned into canals to irrigate desert crops or pumped into that gaping maw, that spreading blight, that bloated mass that is the Phoenix metropolitan area. Phoenix itself will continue to gobble up no longer arable farmland until it fills half the area of the state.

§

My status on this voyage has plummeted. The hardy river runner has been reduced to ballast. Bow scum, an ape without an oar. Bucking the steady headwind with the broken paddle at his feet, Tom works hard under the blistering sun. He eyes the back of my head and quietly contemplates homicide. River runner my ass, he thinks. And this absurd expedition was definitely his idea. Murder is certainly justifiable, Tom reasons, you arrogant little nearsighted bastard, but what if I have to portage?

The brilliant sunlight reflects off the surface of the river and from the shiny metal bottom of the canoe, slowly cooking us under our hats. The hot wind acts like a natural blow dryer set to high. We use an old desert trick to keep cool, repeatedly dipping our shirts and hats in the water during the long afternoon. The evaporative effect is incredible. Putting on a wet shirt raises goose bumps along our arms and for a short time our heads and upper bodies are actually cold. The air is so nearly without moisture that the wind dries our garments within minutes.

The river narrows and picks up speed as the canyon walls creep closer. Tom is very tired and I am unknowingly suffering from a concussion and heatstroke. In the intimate way that only old friends can, we snarl at each other as the ill-fated canoe bangs off one big sunken rock after another. The river suddenly narrows into a tight, rushing chute that drops several feet over half-submerged gray and black boulders. Tom strains to get the canoe to shore but the current is too strong. I jump out, get my feet on a big flat rock, and between Tom's inspired paddling and me wrestling with the bucking canoe we manage to get it beached before we go over the drop sideways.

Walking up and down the river to survey the situation, we stretch our cramped legs and sip hot water from canteens. The chute ends in a dark pool, which empties out into a wicked maze of canoe-busting boulders.

"We're gonna have to portage, Stick," Tom announces. "We can't run that."

Portage: the carrying of boats and gear overland from one body of water to another, or to another point on the same body of water. Up to this point in my young life I had read many books on exploration by boat and had foolishly created a romantic image of canoe voyages. Part of this naive view was a vision of a canoe portage as carrying a lightweight birch bark canoe along a path in a green forest with my merry companions chanting happy songs. My fantasy is shattered while carrying our massive boat through cactus and mesquite thickets, my gloomy companion and I hurling creative insults at one another. It takes three trips up and down the brush-

choked river to complete the portage. Our legs are scratched and bleeding from thorns and deerfly bites.

Shoving off once again, we drift slowly down a welcome stretch of flat water. Tom fishes a bottle of aspirin from his pack and shakes out half a handful.

"Share?" I ask. We've both got throbbing headaches from the sun and glare. He takes three, contemplates dumping the rest overboard, but finally hands me the bottle.

"Maybe we're out of the worst of it. It's got to open up soon," I say, looking up hopefully at the steep canyon walls. When planning this trip, we had bravely decided not to take any maps, because after all, John Wesley Powell didn't have any maps, or at least maps of any accuracy, when he and his men became the first to navigate the Colorado River through the Grand Canyon.

When the sun finally drops below the rim of the deep canyon the long shadows bring welcome relief, but the murky water is suddenly hard to read. Long stretches of navigable water have been rare, and it is during these times that Tom can lean on his paddle and get the canoe up to speed. We are running fast down one of these stretches, assisted by a swift current through some minor white water. Without warning, we abruptly run up onto a piano-sized rock sitting squarely in the middle of the river, a mere six inches under the surface. With our gear stuffed under the stern seat and the bow storage, all the weight the canoe carries is at each end. The bottom of the canoe just caves in, folding up with a sickening crunch as we come to a jarring stop on the rock. I let my momentum carry me out of the canoe, put both feet down on the rock, and wrench the canoe free. It still floats. The keel is broken, folded up with a deep crease that is perpendicular to the long axis of the boat. The bow and stern droop sadly into the water.

And yet my stalwart companion continues to do a fine job with the now reconfigured canoe. Avoiding all the rocks is out of the question; there are too many. No amount of fine paddling can move a boat through shallow rocky water. It is here that I finally earn my keep, spending half my time in the river pulling the canoe over and around boulders and sunken trees, the canoe wading reluctantly behind me like a tired mule. It is not a task that requires skill, just brute force and determination. Once, I misjudge the current, and the canoe comes right over the top of me, slamming me against the river bottom. After a time, perhaps due to the effects of heatstroke, the canoe seems to stop moving and it is the river that comes at us, an endless

highway of rough water and sharp rock, no stretch just like the one before, each offering some new danger to the rapidly disintegrating canoe.

By the end of the day the sun has pounded me into a stupor. Evidently, we stopped to investigate an old mine and paused for a rest in a magnificent grove of old mesquite trees, but I to this day I can't remember these things (and to this day Tom has not showed me the "very nice photos" he took during our run through "The Box." I think he's still upset).

As darkness falls the canyon finally begins to open up again and we find a nice sandy beach backed by a thick stand of mesquite trees. It's perfect. There is plenty of wood, the fine old thicket of mesquites for shelter, and a plethora of big red ants. Two huge ant mounds are visible underneath the trees and the sand is swarming with neurotic little soldiers, each packing a poisonous sting.

Outnumbered a million to two, we retreat to a sand bar in the middle of the river. No ants. No wood. No shelter. Tom eats a cold meal, but I have no appetite and little energy. I spend the evening just sitting in the remains of the canoe, listening to the river. The cool breeze off the water is like ice on a hot, bruised ankle; it makes the disastrous day worthwhile. A nighthawk booms overhead, picking off the insects hovering over the water. Something splashes downstream, probably a fish, and a lone coyote screams at the darkening sky. There is nothing quite like camping along an unspoiled length of desert river, especially during the hot months. It is like being held gently in the arms of the earth.

Having lived almost all of my life in the Gila drainage, I have become somewhat attached to the river. I'm an Arizona boy, and I spend as much time hiking, exploring and camping as I can. I have spent much more time on the tributaries of the Gila; the Verde, the Blue, the Hassayampa, the San Pedro, the San Francisco, and of course the mighty Salt than on the Gila herself. In the aggregate, I've spent months backpacking and day hiking along these small rivers, climbing the steep canyon walls to explore a ruin or to sit and glass for deer and javelina, and then (if its summertime) to plunge into a deep green pool. The times spent on these little desert rivers have been some of the best times of my life, sort of a combination of coming home and going to heaven.

Then there are the sub-tributaries, the creeks, rivulets, washes and rills that feed by the hundreds, the thousands, into the major branches of the Gila River. Most of these are dry part of the year, some only run intermittently, and some only when it rains good and hard for a long while.

These watercourses in the main do not offer anything like the luxuriant plant growth and teeming wildlife of the principal tributaries (although some of the lesser streams that carry water all year, like Aravaipa, are absolutely magnificent), but all of them have something interesting to offer. It might be a cave, an old ruin, spectacular isolation, a hidden pool of good water, or a good half-day spent watching bighorn sheep. Alder Wash, Black River, Ash Creek, Greenbush Draw, Tres Alamos Wash; hundreds of long, sandy fingers reaching thirstily into the dry old hills. Some of these places are so isolated, untrammeled and unknown even today that I don't dare mention them for fear that some imbecile will hear of the place, make the long, difficult trip out there and fling an empty beer can into the canyon. You know what I mean.

"Looks like your dad's canoe picked up a couple of scrapes," I offer, hoping to get a laugh from Tom.

He doesn't think it's funny. "It's ruined. Period. However, I would still say that the canoe looks better than you. You look a little rough."

"I feel really bad. I guess that hit I took on the head this morning gave me a concussion, or something. I can't even move."

"I noticed."

"You need some help?"

"For what? We've got no wood, so no fire. No fire, no hot food. What's there to do?"

Fearing that the river might come up overnight, we leave all our gear in the canoe except our sleeping bags and prepare to sleep on the damp sand. Despite the searing heat of the day, the river has cooled the night air and a delicious breeze flows over the sand bar. This is good. We will be able to sleep inside our bags tonight.

Camping anywhere near a water source in the desert during the hot months often poses a dilemma for those who prefer to sleep under the stars. The nights are usually too hot to wear much in the way of clothing. Burrowing inside a sleeping bag is out of the question. Lying on top your sleeping bag, the view of the stars is often blurred by a cloud of bloodsucking insects maneuvering into formation. The little vampires drive you inside the bag where you lie sweltering in your polyester sauna, soothed by the keening whine of lovely little wings. Overheated, you finally have to unzip and crawl out, offering your sweaty body to the delighted bugs. Back inside you go. It's either give blood or sweat.

No problem with being too hot tonight. Sliding into my bag, I make the wonderful discovery that there is as much sand and water inside my bag as out. Nothing like the sensual feel of cold, slimy, gritty plastic against tired sunburned flesh. Aaaah, yes. This feels so very familiar. All I need now is a large carnivorous insect exploring my toes and a couple of sharp rocks underneath my ribs, and the night will be perfect...

Despite it all, I sleep like the dead and wake up to a beautiful morning. The sky to the east and the surface of the river glow with a fresh orange blush that reminds me of the inside of a cool ripe cantaloupe. The living river looks delicious. Standing up slowly, I carefully peel the sticky fabric of the sleeping bag from where it has stuck to bleeding scrapes on my elbows, knees and ribs. Wading into the river, I sink down slowly into the cool water and wash the sand and blood out of my shaggy hair. Gingerly feeling the knots and scrapes on my head, I stand waist deep in the river and turn a full circle, taking it all in.

The canyon is spectacular; big and rugged, with towering walls of sharp rock that continue to keep the human race at bay. This area is still nearly untamed despite a century's worth of vigorous effort to turn the wilderness into something it isn't. There is no litter and no roads. The river sweeps around our lonely sand bar, bubbling and gurgling its way through the parched desert. Fifty yards away from the river, mesquite and acacia struggle for survival on the rocky talus slopes below the cliffs. The bounty of the river does not reach them. They could be fifty miles out on a dry plateau for all the good the river does them. In wild contrast to the envious desert above, the banks of the river are green and lush with water-loving plants.

Massive cottonwood trees four and five feet thick at the base lean out over the calm water. These big trees can gulp fifty or more gallons of water a day, their huge branches thick with glossy green leaves which carelessly surrender the water to the hot dry air. The cottonwoods rise to eighty, a hundred feet tall, towering over the stunted mesquites rooted deeply in the hard ground. In midsummer on these barren slopes over the river, the temperature near the ground reaches a hundred and twenty, a hundred and thirty, a hundred and forty degrees, day after day after day. The desert plants are getting by in a year on less than what the cottonwoods use in a day.

Alternately wading and floating downstream, I quietly drift up on a herd of javelina taking their morning drink. Spooked, the pigs snort and charge off downstream in a spray of mud and water. They in turn disturb a great blue heron fishing off the riverbank, and the big bird does a fine imitation of a miniature pterodactyl as it circles overhead. The river has an ancient and primal feel to it this morning, and the raw, savage canyon frames it perfectly. When Tom and I make the Big Trip, we'll camp here for two weeks.

We eat a cold breakfast, stow the gear securely in the canoe (learned a few things yesterday!), and are on the river before the sun comes over the canyon walls. To our surprise, a high cloud cover forms soon after sunrise and the morning stays pleasantly cool. The river is much more tranquil, but the wobbly canoe gives us some exciting moments, including one where the current wedges us against a big snag that is draped with barbed wire. After an hour of moderate work we exit the mountains and the river begins its placid run through desert and farmland. To our surprise and delight the canoe does not sink, and by noon we reach the place where we had parked my truck a couple of days earlier.

Resting in the bed of the truck with its bottom to the sky, we finally see the canoe's full scope of damage. The canoe looks as if we have beaten it with sledgehammers, filled it full of big rocks, and tossed it out the back of a moving pickup at about seventy miles an hour. I never did have to move to Montana, however. Soon after our voyage someone bought that canoe (and whoever it was must have been a fool, an artist, or a body-and-fender hobbyist) and Tom, Sr. never was overly upset about the damage done.

The voyage complete, I came to have a new respect and appreciation for the Gila. I forgave in full the San Francisco River, and came to camp on her sandy beaches again and again (and Tom, to my complete astonishment, remains my stalwart friend to this day, although the subject of canoes does come up more often than one might expect in routine conversation).

There is hope for these little rivers, more than anyone reasonably would have expected looking ahead from our 1981 canoe trip perspective. Many people outside the community of dedicated conservationists and the cognoscenti of environmental literature are becoming conscious of the incredible amount of damage that has been wrought on the American west via gigantic water projects, cattle ranching and large-scale farming of water-craving plants. Certainly, Mark Reisner's slashing "Cadillac Desert" has opened a lot of eyes. Increasingly, people are wondering why nearly every river in the West has at least one dam on it, and some are actually demanding the systematic removal of dams. Glen Canyon dam, which forms Lake Powell, increasingly seems to be one dam too many on the Colorado River. Sound crazy? Perhaps it does, but so did the Endangered Species Act when it was first proposed.

Regardless of how one views the wilderness, whether it is just a big empty place to throw beer cans on the weekend, or even if you are one of us cactus-hugging fringe neurotics who actually get emotionally attached to rivers and such, the land remains awesomely indifferent to whatever we do. What a place this must have been in the 1850s! Even today the half-strangled Gila can still give you a tough go of it. I was humbled.

II. GILA WILDERNESS

It is six o'clock on a Monday morning, and for the first time in longer than I care to admit, I am truly a free man. Instead of dress slacks, shirt, and tie, I am wearing faded camouflage pants, a ragged green shirt, and well-worn boots. Hanging onto my head like a tired old badger is my favorite hat, a tattered, malodorous and absolutely offensive piece of headgear.

I am not going to work! It hits me again and again, the sweet taste of liberation. Life has smiled on me. There is no briefcase full of engineering notes beside me. Instead, a sleeping dog, and packed into the camper shell behind us are all the accessories of the Free Man (and Free Dog): tent, sleeping bag, backpack, daypack, water, and food.

While rolling eastward on the interstate, past the glass towers of downtown Phoenix glowing with the rising sun, I recall years of sitting in my tiny cubicle, a drawing, tolerance study, or some other technical document spread across my desk. I remember staring at the pile on my desk, attempting to concentrate, avoiding the urge to dream. Trying not to fantasize of a lonely highway crossing a vast desert, of climbing a rugged canyon wall with my heart pounding as I search for a route through the rimrock; of the sight of a pack of hunting coyotes outlined against fresh snow and working a draw with fluid precision.

Ah, this moment of sweet escape, tearing along the freeway with the office herd, all tucked into their hurtling steel and glass cocoons. The little truck strains to keep pace. The traffic thins as office workers stream off the freeway into the inner city. Smiling, I keep to the center lane.

I continue east through the sprawling thirty or so miles of Tempe, Mesa, and Apache Junction, and on past the smog-wreathed battlements of the Superstition Mountains, brooding over the subdivisions. Finally, after an hour of driving, we clear the city and roll out onto the desert. Smog thins. Traffic slows. I pull over, and the dog and I traipse through the bushes, sniffing and stretching and looking for something defenseless to piss on.

Onward across the beleaguered expanse of Sonoran Desert between Apache Junction and Superior, all destined for subdivision, rape, and pillage. Unaware of their fate, the doomed saguaros stand proudly on rugged slopes.

I stop for coffee in Superior. It is early in the morning and already hot; it feels like ninety or ninety-five. Chrissie and I sit in the shade at the edge of the cracked asphalt parking lot and admire the big cliffs that dominate the little town. Actually, I admire the cliffs, while Chrissie patiently tries to bum some of my candy bar. The resolve of the canine family is admirable; to my knowledge this dog has not received a piece of candy in her entire life, but she never abandons the quest.

The west face of the cliffs is in shadow. Somewhere up there is the infamous Apache Leap, where Apache people, unwilling to be captured by U.S. Dragoons during the bleak Indian War days of territorial Arizona, jumped off the cliffs to their deaths. Or so the legend says.

Before we leave town, I cross the main highway and drive through downtown Superior, where the heyday of the big copper mines seems to be a thing of the past. Many of the buildings are abandoned and boarded up, and the town has a half-empty feel to it, but I am reluctant to leave. It's just a little patch of town on the edge of all that desert, but it's my kind of place. It reminds me of the Bisbee of my youth.

Bisbee was one of a long series of small Arizona towns that my family lived in when I was growing up. Unlike big cities, where the adjoining land has been abused for so long that it looks like an extended vacant lot, the desert or forest comes right up to a small town. No matter where we lived, we somehow managed to have a house right at the town limits. As a young boy, I was allowed a strange sort of liberty. I was not permitted to cross any streets alone, but it was fine for me to trek into the boonies for as long as I wished. It is deeply satisfying to be able to look out your bedroom window and plan your next expedition.

The truck whines and wheezes over the steep mountain highway into Miami and Globe, small desert towns baking in the sun. East of Globe the land becomes empty. Finally far enough away from the smog factory of the Phoenix metropolitan area, the sky turns from dirty white to a crystal blue. The highway runs eastward through the classic range and basin country of southeastern Arizona; huge expanses of sun and sky decorated by long chains of tall craggy mountains.

These mountain ranges, due to their great size and also to their physical separation from other mountainous areas, have been called "sky islands." They have also been likened to great ships sailing on a vast desert sea. Scientists have discovered species on these "sky islands" that are unique to any other on the planet (I was offended until I learned that they were talking about animals and insects. I thought they might have been referring to my hiking companions).

These sky islands are some of my favorite places to hike. If you are in good physical condition, you can, in a day's walk, begin your hike in hot and dry desert country, pass through several different plant and animal communities during the ascent, and lunch in a cool and damp spruce forest. As a rough approximation, climbing a thousand feet of mountainside is equivalent, weather and flora-wise, to traveling north three hundred miles. At an elevation of ten thousand feet, which you would approach or exceed near the summits of many of these southeastern Arizona mountain ranges, you would enjoy your lunch in the shade of plants similar to those found at low elevations along the Canadian border. When considering the return trip, one of these sky island hikes can total an altitude change of eight to ten thousand feet.

A few miles east of Peridot, I stop to give the dog a break. The sun is straight overhead, and heat waves shimmer off the highway. To the south, the remote summit of eight-thousand-foot Mount Turnbull is a mouth-watering cool blue. We cross a barbed wire fence and stroll through the desert. Like a lot of areas in the dry southwest, the land is seriously overgrazed. The strip of ground near the highway is mute testimony to the hard years the land has seen. Off limits to cattle, the grass is thick and two feet tall. Inside the fence the ground is raw, exposed and eroded, and thick stands of scrub mesquite, catclaw, and prickly pear have taken over.

Ranching is a time-honored tradition in the West, a holdover from the frontier days, the wild days. I look across this tired land, hammered by a century of people and cattle and the eternal drought, and wonder about tradition. Besides, I pity the cow that has to earn a living out here.

We meander back to the truck, Chrissie panting hard in the noonday heat. She looks at me gratefully when cool air starts coming out of the vents. Turning on the air conditioning is but one of many miraculous acts that humans can perform for their dogs. I am convinced that Chrissie thinks that I am a mighty wizard, in command of forces unseen. Consider the twice-daily miracle of The Food, for instance. The mighty wizard, beyond all hope, produces a bowl of tasty food when before there was naught but an empty bowl. Twice a day my dog is amazed and overjoyed. Chrissie puts her head in my lap and enjoys the miracle of cool air.

After passing through Safford, we drive north to Clifton. I dearly love Clifton. She is true to her roots. She sits rugged and pretty like a true-hearted old-west madam, right astraddle of the most perfect little desert river you ever saw, a weathered town surrounded by shattered cliffs and old mine workings.

The truck clatters over the bridge and I turn north along the well-kept but lightly used dirt road that runs along the east bank of the San Francisco River. Several houses gutted by the big flood of '83 are sitting empty like old hollow-eyed ghosts, decaying in the hot sun. The river, not much more than a middling creek most of the time, became a giant overnight and smashed the town. That's Arizona for you. No rain for weeks, or months, and then you get a year's supply in two days.

The wide road narrows when it begins to cross the river. I shift into four-wheel drive to take the deep fords. The truck claws and bucks, the fat knobby tires kicking up big rocks that crash into her underside. Dripping, she climbs out of the water time and again.

I have come to greatly appreciate four-wheel drive. My father was an avid outdoorsman, a four-season hunter and fisherman. To him, inclement weather was a minor inconvenience, and staying home because of cold or snow or rain was to miss out on half of nature's glory. We lived in the mountains for most of my youth, and my father maintained that the weather was good about half the time, and the other half ranged from miserable to awful. So, he went. And with a son being sort of a built-in friend, I went with him. After a few years of determined sniveling I actually came to appreciate the power of a driving storm, the serene beauty of an icy morning, and the lush feel of a summer rain. I still whined about the misery of it all (and still do), but I went, and enjoyed myself.

The downside of experiencing the outdoors with my father was that he was a stubborn two-wheel drive man. He was fond of maintaining that four-wheel drive vehicles were expensive toys, and all they were good for was getting you deeper into the boonies before you got stuck. He was also fond of digging, cursing, and walking considerable distances while wet and muddy. He must have been. It was an activity we engaged in on a regular basis. By the time my youth had passed I was expert with shovel and winch.

Fifty yards from the river stands my favorite sycamore tree, an ancient, twisted giant that offers a refuge of deep shade, enough for the truck and a generous campsite. When I switch off the hot motor, the silence comes rushing in. Digging my cheap lawn chair from the camper shell, I sit with my toes in the sand, eyes closed. My senses slowly adjust from roar of engine, smell of overheated rubber and gasoline, the glare of chrome and asphalt, to wind in the leaves overhead and the smells of hot rock, foliage, and the river. The road tension flows out of my shoulders and through my feet into the warm sand. Big red ants trek back and forth through the sand. The wind is very hot and very dry. It is desert summer.

Walking along the river, I decide to take a crack at climbing out of the canyon. Within a few minutes, it becomes obvious that the day is too hot for long-haired Chrissie; the sand burns her feet and her eyes turn glassy. I give up on trying to find a route up the cliffs, abandon the idea of encouraging the beast, and finally just lift her off the hot sand and toss her over my shoulder (how humiliating!). Retreating into the cool water, Chrissie and I spend the afternoon in the river. Chrissie splashes around in the pools, snapping at the tiny fish, as I bodysurf down the rapids near camp. This beautiful little river is a jewel in the wasteland, a sinuous oasis in the vast tumbled desert. Floating on my back down a long, deep pool, the view is through thick green boughs, past red cliffs and raw, barren peaks into the intense depth of sky. The combination of color, caress of water, sigh of wind, smell of tree, river, and the dry sweet desert brings on a profound state of tranquility that only primal landscape can bring. The long, hot afternoon gradually gives way to a lovely evening cooled by a thunderstorm.

While playing in the rapids, I somehow lost my wedding ring. It just slipped off my cold, wet finger sometime during the long afternoon. I'm in deep trouble. That was my second wedding ring, and I'm still on the same wife.

As the sun goes down it becomes so quiet that the dog and I, sitting on top of one another, scarcely breathe. The sun bursts free from the heavy clouds for just an instant before it drops behind the ragged cliff at our backs. Every leaf on the gnarled old sycamore across the river burns with green flame against the shattered red rock of the canyon. The moment is so pure and sweet that it hurts.

§

Up before dawn, I pack and load camp and dog. Chrissie won't eat. Back down the river and through Clifton I go (vowing to always return).

Highway 75 catches my eye on the map, and we tour southeast through Duncan, Arizona and Virden, New Mexico, crossing the Gila River over and back again. I am amazed, as I have been many times, at the contrast between the lush land near the river and the stark simplicity of the desert. The desert here goes on forever. Canyons rarely cut the immense empty flats, and the view of the craggy New Mexico mountains is clean and straight for fifty miles. Ah, the picturesque and forsaken desert of southern New Mexico, empty and forgotten. The mountains look absolutely devoid of life (an illusion).

After a couple of hours, the highway intersects I-10 at Lordsburg. An hour later, I stop at my favorite rest area along the freeway. The rest stop has a name, an official name on an official plaque, but I always have thought of this place as the Rest Stop of the Wind. I have to hold on to the truck doors tightly as I open and close them. When getting back into the cab, I keep my foot carefully against the door in case the wind shifts and whips the door against my leg. In the winter, it is a cold wind that never ceases. In the spring and fall, it's a cool wind that never ceases. It's now high summer, and the dog and I lean into a hot wind that never ceases.

The wind blows Chrissie's hair into weird shapes that change as she turns. With the wind at her back, she looks like a small black and white lion with a fierce mane. When the wind catches her from the side, it whips the hair up into a full-body Mohawk that makes her look twice her size. From the front, the wind pins her ears back and molds her hair flat against her sleek, muscular body. She looks like she's flying, my amazing shape-shifting dog.

Except for the wind, it is a typical rest stop. I use the men's room, and after washing my hands, decide to dry my hands on my pants rather than use the ridiculous hand dryer. Think of all the money they save on paper towels. Why buy paper towels, when people can use their clothes instead? Outside, I try to drink from the outdoor fountain, and get mostly wet. The incessant wind blows the arcing fountain of water into a fine spray. By this time, the dog is irritated by what the wind is doing to her fur. She lies in the lee of the truck with her head tucked between her paws.

From Deming, my impromptu route takes me north on Highway 180 across fragile desert grassland. We continue on Highway 61, which sweeps northward along the Mimbres River. We pass through Haywood, New Mexico (population fifty, including pigs and chickens). The country becomes more hospitable (which does not necessarily make it better) as I cruise north. This is open range and basin grassland, cut by the verdant swath of the riparian ecosystem of the river bottom. Purple mountain's majesty dominates the horizon to the east and north. Continuing north through the town of Mimbres, the drive winds up through piney dry desert forest, very like my homeland White Mountains. New Mexico is so much like Arizona. They are sister states, at least geographically. New Mexico is a bit less rugged and beautiful, but is also less populated.

When I am nearly to my destination of the Gila River Wilderness, it occurs to me that I haven't yet called my wife as I had promised. Being so far from a phone I am tempted to let it go, but guilt over losing the ring sends me down the road. I drive from the end of the highway south to Silver City, call home (got the accursed answering machine!), then back north to the Wilderness. It is an exhausting drive along the lonely, wretched, cramped little highway. The road is populated by deer, an occasional nervous tourist, local maniacs in giant four-wheel drive trucks (baby tucked in tight, cheek to cheek, Hank Williams on the box) and anonymous weirdoes in ancient Batmobile Cadillacs.

Tired and strung out from too much time behind the wheel, I get lucky and get the prime campsite on the West Fork of the Gila River. I back the truck up under a tree right on the riverbank.

We take a short tough hike up a rocky ridge near the creek. Chrissie is bored, refuses to walk with me, and does whatever the hell she pleases. Fine. Her ears will recover when I call her for dinner. The soaring, fluted cliffs on the far side of the creek reflect the warm glow of the setting sun in a most soothing way, and I sit on the ridge until the sun goes down, absorbing nature's blush.

After setting up my camp and wolfing down some canned stew, I present Chrissie with a luscious bowl of dry dog food (much more palatable than my stew). She still won't eat. This sudden lack of appetite is very unusual for this dog, as she is always ready to eat anything edible and a variety of things that I consider to be clearly lethal.

After dinner, we take a walk along the river in the pitch-black night. Careful to maintain my night vision, I keep my flashlight off. My hopes, however optimistic, of catching a deer or raccoon in the quick beam of the flashlight are crushed by my enthusiastic dog, which makes more noise than a bull moose charging through a willow thicket. That's all right with me, though, even when she speeds by and wipes a gallon of cold water off on my leg. It gives me great joy to see her set free from her concrete environs, running hard, going wild with all those feral smells everywhere she turns. I have always enjoyed the company of dogs, and this one in particular. There is something about those warm, brown eyes and the ready companionship that makes me feel not alone when I travel with a dog. And besides, this dog can go straight up a granite cliff like a big black-and-white spider, can march all day on a handful of dry food, and has no fear of anything, except being abandoned.

My love is put to the test soon after, for, in the deepest, darkest hours of night, my stanch companion scares the living hell out of me. Chrissie is tied to the rear bumper, and for no apparent reason runs full speed out to the end of the rope, making a terrible noise and jolting the pickup. For a few seconds I am convinced that Jack the Ripper is coming through the door of the camper shell, and immediately take appropriate self-defense measures, which include bashing my head against the fender well and cursing in two languages. I yell at the stupid dog for a while. She has already forgotten why she was running around, immediately forgives me for yelling at her, and goes right to sleep. I do not.

The dim light of dawn blossoms into a beautiful mountain morning, cool and clear, full of mystery and promise. The first golden light of the day comes gently down on one of the last relatively pure areas on the planet. The Gila Wilderness, designated as such in 1924, was the world's first wilderness area.

On the way to the cliff dwellings, I stop in at the ranger station and talk with the Ranger for a bit. Nice man, and I envy his job, mainly because he obviously doesn't have much to do. He is friendly, relaxed, and welcomes visitors simply for the chance to talk.

This is very encouraging. If a true wilderness were managed as such, there wouldn't be a damned thing to do except to go outside and watch the clouds. On the other hand, if the land is being logged, mined, gouged, maimed, reamed, inflamed, corrected, flattened or otherwise improved, then there is enough work to keep a hive of bureaucrats busy.

At the cliff dwellings, which are the centerpiece of the National Monument, a sign greets us. Amongst the Multitude of Prohibited Things is: No Dogs. I park the truck in deep cool shade and Chrissie guards the vehicle while I take the short walk up to the cliff dwellings. Built by people of the Mongollon civilization in the late thirteenth century, they are very impressive ruins; stone and mud buildings tucked into shallow caves under precipitous cliffs. The view from the ruins is down into a pretty little canyon.

I always feel like I'm trespassing when I find a ruin somewhere in the backcountry, back in some side canyon, five hundred vertical feet up from the watercourse. I'm reluctant to enter and usually don't poke around much. It's not my house. No one invited me to come.

Most of the sites that have been developed for tourist viewing don't strike me the same way; maybe it's the paved trail or the little informational signs, or just knowing that archeologists have sifted every grain of sand within a hundred yards of the place. The ghosts have all been chased away; the memories hauled off and put in a box in a museum's back room.

This is different. There is a nice easy paved trail and all the little signs you would expect. Perhaps it's because I am alone here this morning, or that it rained last night and everything looks new, but these ruins haunt me. It is a poignant experience to see where people once lived in such a simple manner. That was a long time ago, and they are gone, taking most of their secrets with them.

The place feels comfortable. Across the canyon is a series of small flat areas where the people dry-farmed, and many of the small creeks in the nearby canyons run all year. Game is plentiful here now, and it must have been like a supermarket back then.

A bit tougher than a supermarket, actually. The meat didn't come wrapped in plastic and sit waiting on the shelf. It came wrapped in tough hide, and sometimes had big teeth, and didn't wait at all. The average human life span back then was between twenty-five and forty years. My thirty-five years would have made me an old, old man.

Chrissie is happy to see me. She always is: "It's that same guy! I am saved! I won't have to die in the cab of this truck!"

The West Fork of the Gila River flows near the ruins, and we spend the day hiking along the water. The country is very remote and lovely, and the dog goes berserk with the wonder of it all, running all day with the wilderness on her nose. We tear through the little river, soaked and muddy, and crash through thickets and scare a lot of wild animals from their resting places.

Chrissie nearly bags a turkey with a magnificent, hovering leap. It is an image I will never forget. The early morning sun catches the water droplets on feathers and fur as they explode from the brush twenty yards in front of me. The turkey is gobbling in fright, and the dog gives a guttural growl as she jumps after the rising bird. Her lean black and white body stretches full length a good five feet above the riverbank, jaws open, paws splayed out for balance, ears pinned back. The big bird strains hard to grab more air and the two just seem to hang in space. Her jaws snap together on tail feathers, and the turkey accelerates away as the dog spins down, down, down into the river.

Chrissie. (Illustration by Jan York)

I bash my knee on a rock. Hug a tree. Inspect a lizard. Look for elk. Collect pine needles. Watch vultures circle. Listen to the croaking of ravens. Watch a rabbit hide from the dog. Try to imitate a birdcall. Taste pine sap. Chrissie digs a big hole, and proudly sits in it as I praise her for her fine work. Dog eats a big bug. I chew on weeds as we trot across a big flat mesa, stumbling over the rocks and sweating like a river, reveling in the soaring happiness of just being Out There. We lie in the shade of big Ponderosa pines, Chrissie draped across my legs. She bums part of my sandwich. A large, unidentified flying insect stings the living hell out of my arm, and then escapes before I can whack it. We lie on a rock ledge and watch the thunderstorms build in the east. Dog chases a squirrel. I throw large rocks off very high places. Chrissie lies in the creek, smiling.

Near the end of the day, we lie unmoving in the hot sun, sweating and panting, as a group of deer moves just ahead, unaware of our presence.

Gripping one of her rear feet, I whisper urgently in her ear, "Stay. Stay. Stay." Eyes bugged out, she pulls her foot free, and I drape most of my one hundred eighty-five pounds across her back. "Stay, Chrissie! Let me get a photo! Stay!"

She squirms free and tears off in high pursuit. The deer stop, startled, and then with the absolute disdain that truly fine athletes have for lesser creatures, effortlessly leave her behind, going straight up a steep hillside, gray and brown, flashing and flowing through the trees. A glimpse of an improbably muscled flank, flexed for leaping...gone. Dog drags it back, humble and seriously outclassed.

We climb the steep canyon walls looking for the perfect ledge. We find it, and just sit, listening, listening...

Far below, the river makes the ancient music.

BIBLIOGRAPHY

Bourke, John G. (Captain Third Cavalry, U.S.A.). *On the Border with Crook*. Charles Scribner's Sons, New York, 1891.

Cordell, Linda S. *Prehistory of the Southwest*. Academic Press, Orlando, FL, 1984.

Corle, Edwin. *The Gila, River of the Southwest*. Holt, Rinehart and Winston, Inc., 1951.

Crowley, Kate and Link, Mike. *The Sky Islands of Southeast Arizona*. Voyageur Press, Stillwater, MN, 1989.

Faulk, Odie B. *Arizona, a Short History*. University of Oklahoma Press, Lincoln, Nebraska, 1970.

McNamee, Gregory. *Gila: The Life and Death of an American River*. University of New Mexico Press, Albuquerque, 1994.

Reisner, Marc. *Cadillac Desert*. Penguin Books, New York, New York, 1986.

Salmon, M.H. *Gila Descending: A Southwestern Journey*. High-Lonesome Books, Black Range Station, San Lorenzo, New Mexico, 1986.

Webb, George. *A Pima Remembers*. University of Arizona Press, Tucson, 1959.

Worchester, Donald E. *The Apaches, Eagles of the Southwest*. University of Oklahoma Press, Norman, 1979.

Lava and Sunlight

Struggling desperately with the tuning knob on my obsolete radio, I escape from the urban horrors of Phoenix, weaving the required mad path through the overheated concrete maze. The truck punches a temporary hole in the pall of noxious gases that lie over the city like a grubby wet blanket. Chanting my personal mantra, I run the gauntlet of Apache Junction, dodging motor homes and luxury sedans uncertainly piloted by octogenarians from Ohio: "Save Arizona, move back East!" "Save Arizona, move back East!" They can't hear me, the blue-haired, white-knuckled (35 mph in the fast lane) relics. Emerging at last onto open highway, I mash the throttle to the mat and push the tiny engine up to maximum rpm, a white blur through the small Arizona towns of Superior, Show Low, and Springerville. Freeeeeedom!

The battle of the airwaves continues as I drive north: no tape deck, no CD player. It's radio or the whine of off-road tires. I eventually lose the war and reluctantly yield to the banality of rural radio. Call-in swap meets and country music. Commercials for tire shops and feed stores; long, crackling silences while the part-time DJ fumbles for the next tune:

> "Ah sure am blue,
> She was untrue..."

We don't listen to any of that Rock n' Roll stuff out here, boy. Between Show Low and Springerville, rolling across the stark volcanic wilderness of the northern reaches of the White Mountains, I pick up a heavy metal song on a crackling, fading AM station. Frantic to hear anything but the twanging, bleating, sobbing sounds of country music, I crank it up until the speakers distort. One last chance to bang my head on the steering wheel;

when the station fades to white noise it is time to give up. The tires have a soothing whine, anyway.

My general destination is northwestern New Mexico. Somewhere out there is a big lava flow I heard about many years ago. The only lava flow I have explored to any extent is the flow that spills down into the Grand Canyon from the old cinder cone of Vulcan's Throne. That was a tough hike. It was hotter than hell, very steep (the Canyon has a tendency to be that way), and my nice new boots were rags on my feet when I finally got out of there. About all I know about lava flows is that they were once molten rock and are tough on boots.

So, the lava being my general destination, I'm not precisely sure where I will end up. And who cares? I've got everything I need here in the truck; ample food and water to live comfortably for a week, enough to survive on for two weeks or a bit more if I'm crafty. One of my favorite leisure activities is to load my truck with outdoor gear, food and water (companion optional) and hit the road, stopping when I please. When evening falls I usually camp in remote, non-designated campsites. The objective is to avoid designated campsites, where you will find government officials, pavement, people, motor homes, fees, and other things best avoided. If the area is interesting, a hike into the country surrounding my camp for an hour, or maybe a day or two, is the most satisfying and, in fact, the only way to know the terrain. After a fine night under the stars it's back to the pavement, heading in the original direction, or not. A general plan is fine, but getting too specific can ruin the experience. The point is to drift until something exceptional turns up.

Enough open, wild land to choose from is one of the reasons I love living in the American West. Having four-wheel drive gives me that much more freedom. I don't have to stay on the pavement, or even a maintained dirt road. A rough little two-track leading into a craggy canyon can be very inviting, and I can go as far as the road and the truck will allow.

§

It is gentle on the eyes to return here to my boyhood homeland of the White Mountains. A beautiful area, full of wonderful and terrible memories: Green's Peak, Mesa Redonda, Mount Baldy, Mount Ord. The Escudilla. Little Mormon Lake. Salt River Canyon, Black River, Sheep Cienega. Still wild after all these years. No matter how far I go, no matter what I do with this life, no day will pass without thought of this blue-green high country.

From Springerville, Highway 260 takes me south to Alpine, Arizona, and then Highway 180 winds south and east into New Mexico through Luna and Reserve. These little towns are dots on the map. Wide spots in the road. Don't blink or you'll miss 'em. They won't miss you, however. Keep movin', stranger. Drop some cash in the till, but just keep movin'.

In Luna, I stop at what appears to be the only store in town, which by default makes it a General Store ("We're generally open, but not always"). I get out of the truck and stretch, hobbling around the small parking lot in the peculiar Motorist's Shuffle that comes of too much sitting in one position. My left leg feels six inches longer than the right, so I tack back and forth across the lot until I can look both directions along the highway. The town is about a hundred yards long, and from what I can see, this store is the only business open. After both legs are roughly the same length again, I go inside for coffee and a snack.

While I browse for jawbreakers, the proprietor and the only other customer in the place lean on a glass case presided over by a rack of elk antlers and grumble about the "city governments" of Phoenix and Albuquerque. I drift closer and eavesdrop on their conversation while considering the Mesquite Beef Jerky.

"...trying to push city values and city laws on rural communities."

"...situation is completely different here."

"...gettin' damned sick and tired of it. What we need are two sets of laws..."

"...Oh, yeah! That's all we need, more laws! Problem is, we have too many laws..."

I can sympathize. Living in isolated western towns is to step back in time a generation, or two, or three, and that's why people live out here in the first place.

Sympathizing with his price of beef jerky, however, is something else again. If beef jerky prices continue to rise, in a few years, jerky will replace gold as a market standard: "On Wall Street today, investors reacted to the sharp drop in beef jerky prices by..."

I buy some of the Mesquite Jerky anyway, a vat of black coffee, and punish my gut for the next thirty miles with the combination of spiced rope and battery acid. The highway loops and curves through wild, heavily timbered mountains. This is high country, six, seven, eight thousand feet. Between sips of battery acid, I crane my neck back and forth in a futile effort to see it all. The road leaps a deep rocky canyon, and the quick vertical glimpse into its green depths is tantalizing; the slender tumbling ribbon of water, black leaping walls of rock stained with lichen, the dark forest...gone.

You can't really see anything from inside a car. The speed is too great, and the glass and steel keep your senses from gathering the data required to fully appreciate the surroundings your body is hurtling through. But I keep right on driving even though my shoulders are aching. The forest opens onto a gorgeous clearing, a picture-book meadow. Slowing, I peer into the deep shadow of the trees at the far side of the clearing, hoping to see elk. I should stop and walk across that meadow, and smell, and even possibly identify, those purple flowers. If I was walking, I would probably see an elk. If I was walking, I could smell the wet grass and the fresh wind sweeping down through the forest. The highway has me again.

The urge to keep driving on and on is a sort of madness. To step into an automobile and accelerate to highway speeds is to change one's frame of reference so profoundly that a person loses touch with reality to some degree. What is outside the car is the reason I take these journeys, but the hypnotic flow of the asphalt ribbon, the blur of tree and shrub and slum and suburb seize me, and on and on I go.

Just north of Reserve, I break the highway's spell by wrenching the truck off the pavement onto a nicely maintained dirt road. Before that road can hypnotize me as well, I wisely roll down the shoulder and into a dry sandy wash. The rear tires dig in instantly and the truck shudders to a halt. Shifting into four-wheel drive, I tool down the wash a couple of hundred yards.

The backpack and daypack are always loaded and ready. Before locking the truck, I toss in a little food, make sure the slicker is in the bottom compartment of the daypack, and go out for a ramble across the country.

It is a unique landscape of orange sandstone bluffs and sparse desert pines. The bluffs are cut thousands of ways by quiet, secluded little canyons. The steep crumbling walls are bedecked with pinyon pines, one of the most beautiful trees anywhere. Not being particularly energetic, I spend several hours wandering the mesa tops admiring these small bonsai-like pines. Some of them are so perfect, standing daintily within natural pockets of sandstone, that they seem manicured. There is plenty of elk sign about, and I sit under the rimrock and glass with the big binoculars, but I can find no elk.

Refreshed, I roll on, and Highway 32 takes me north from Apache Creek. The country is full of pretty grassy-bottomed canyons, the watercourses lined with big trees backed with sheer walls of sandstone and limestone. This is prime horse and cattle ranch country, and what few vehicles I pass on the road are pickup trucks, many pulling horse trailers. The traffic turns from sparse to nonexistent as the highway leaves the canyons to run straight and level across a volcanic prairie. After five miles, I can see five miles straight behind me in the mirrors and five miles straight ahead. On impulse, I stop the truck, park it just off the road, and perform a martial arts kata squarely on the yellow line in the middle of the highway. As a thunderstorm roars approval, I bow low to the rugged mountains to the east.

Late in the afternoon, somewhere between Apache Creek and Quemado, a rough little dirt road catches my eye, and soon after I make camp behind some ragged, splintered bluffs. Tucked into a juniper thicket, my camp is right on the edge of a big prairie that opens into a huge grassy bowl toward the west. I fuss around camp for a while, then stroll out of the thicket with my lawn chair and binoculars. Placing the chair in front of a small juniper to break up my outline, I sit for an hour or so and glass the prairie, hoping to see some antelope. No antelope, but the view to the west is classic. Some ten (Fifteen? Twenty?) miles to the west, the broad expanse of the grassland slopes steadily into timbered mountains, the myriad ridges running from green to blue to a hazy gray. Behind the main ridge and barely visible through the slanting afternoon rain, the misty view lending secrecy and power to its already great size, the massive bulk of the Escudilla rises to over ten thousand feet high. Between the afternoon thunderstorms that sweep across the grassland from the west, I climb the steep bluffs one by one until it is too dark to see.

Up early, I shuffle to the cab in my underwear, start the engine, and growl down out of the cedar forest in low range four-wheel drive, first gear. I want to be out in the open when the sun rises. I like open places, places where I can see fifty miles.

Parking the truck, I prepare my simple breakfast. Coffee ready and cooling on the tailgate, I don the uniform of the day: camouflage pants, lightweight hiking boots, hat. No shirt. Shirt only when required, like during a gnat attack or when I stop for supplies. Taking my breakfast, I sit in the cab with the doors open and catch the Morning Carnage Report on an Albuquerque station. The delicious aroma of instant coffee wafts through the cab as I eat my sweet roll.

The quiet and beauty of my immediate surroundings is in agreeable contrast with the urban radio report of death, destruction, and graffiti artists rampant. Police desperate. Citizens concerned. Rock n' roll throbs through the speakers nonetheless, which is the real reason I turned on the station in the first place. A young cottontail feeds among the flowers that grow alongside the dirt road. If this were the city, I reflect, gangster rabbits would appear to kick his ass, little blue gangster headbands strapped around their fuzzy ears. Bad rabbits. Armed. Hawks concerned about the sanctity of the food chain...

Gangster Rabbit. (Illustration by Jan York)

As the sun begins to creep toward the horizon, the cliffs above me change color by the second, from black to gray to a delicious yellow gold. Escudilla stands guard on the western skyline, looking like a giant loaf of bread.

After breakfast, I roll my bedroll, pack the lantern, and hurl the rest of my stuff to the front of the camper shell. Coasting down the dirt road in neutral, I pull out onto the blacktop. Radio on and both windows down, I drive north on Highway 32 through increasingly desolate country.

Cruising speed: forty-five miles per hour. Forty-five is the elegant pace of the unhurried person, a velocity suitable for one who has no meetings to attend and no place in particular to be. The wind through the open windows feels good at forty-five, at sixty, a bit harsh. At sixty, you have to wear a hat or the wind makes your hair hurt. The engine chuckles along at forty-five, at sixty she starts to whine. At forty-five, bugs bounce off the windshield, possibly to survive and live full bug lives (but probably not). At sixty, bugs splat. No hope for a splatted bug. Retired people drive at forty-five, to the irritation of busy young people everywhere. Although not retired, as years of work and responsibility loom on my horizon, I am taking an installment of my retirement early.

In Quemado, I stop at the General Store for coffee. Quemado is on the ragged edge of nowhere. Harsh, rugged mountains hang on the southern and eastern skyline, shimmering through the early morning heat waves. Great arid badlands stretch out in all directions with the little town in the middle like an old west movie set.

The side door takes me into the store. Hardwood floor. A battered elk head hangs over the front door. Not exactly top taxidermy; it looks like somebody dug it up and then stuffed it. As I pour my coffee, the powerful young man tending the place moves silently behind the counter, and I quickly decide not to mention the sad elk head. This man may have been the man who shot it. And stuffed it, for that matter. He looks every inch the self-reliant plainsman, like a man who can shoot, skin, eat, and even stuff his own elk. He looks like he could hunt elk with a pocketknife. Coffee paid for, our eyes meet. Level stare. We grunt at each other, deep and guttural, and I leave the store aglow with the thrill of human interaction.

From Quemado, I drive north on Highway 36, then turn east on Highway 117. The desolation turns to moonscape as I approach the enormous lava fields. This is El Malpais, the Bad Land, named by the Spanish explorers who "discovered" this area (Evidently, the Indians did nothing but lay around for thousands of years, only occasionally putting forth a burst of activity to confuse future archeologists). The Spaniards usually skirted the area, not willing to risk themselves and their animals on rock sharp enough to cut, but the Pueblo peoples had been crossing the Malpais for uncounted years. A fierce sun beats down on the tumbled black terrible wasteland, a land of little water and less shade.

Volcanic eruptions have occurred in this area again and again, creating a valley filled with lava. The lava fields are approximately forty miles long and ten miles wide. The spectacular sandstone cliffs that form the eastern boundary of the flows are older by far than the lava below. The sharp contrast between the forbidding, crumpled landscape of the lava and the smooth, towering sandstone bluffs is nowhere more dramatic than at an area called The Narrows. Thousands of years ago, the flowing lava was stopped by this beige and buff wall of rock, and the clever highway engineers have placed the road through the narrow gap between the lava and the cliffs.

At the Narrows I park in the small, well-maintained parking lot. Taking my time, I unlock the camper and prepare my hiking gear: water, food, water, camera, water, survival gear, water, water, water until the pack bulges with canteens. As I work, cars whirl by. Occasionally, one pulls into the parking lot and people get out to troop up and commune with the magnificent sandstone formations, which evidently include a big natural arch. There is a nice, well-marked trail that leads from the paved lot up to the cliffs.

The best way to stay cool in the desert is to keep the sun off your skin. I pull on a cotton shirt, long-sleeved and light-colored, loose fitting for good airflow, and lather sunscreen onto the small remaining amount of exposed flesh. My expensive, state-of-the-art hiking boots will stay in the truck today, replaced by a pair of cheap but comfortable boots normally worn around camp. The lava looks sharp. Just in case the boots come apart, a pair of lightweight, low-top tennis shoes are added to the load. The hat is crimped on low. Good. I shrug on the pack and do a few deep knee bends. The body feels really fine today, limber and strong.

Locking the truck, I turn my back to the cliffs and cross the highway. Few people come this way, into the twisted stone. Although people have been crossing the lava for centuries, and the Pueblos built paths through the lava, I will most likely not find a trail. Climbing the barbed-wire fence, I scramble up onto the lava flow. As far as most of humanity is concerned, this is the south side of the tracks when it comes to natural beauty. Some of us, however, have different tastes. Walking straight north, squarely away from the highway, I am soon into a queer, misshapen world untouched by man. There is nothing that mankind as a species could possibly want of this place, a very positive feature, as it faces no immediate threat from development: "Not a great place for a golf course, eh, Bob?"

I hike all day across the lava, sweat pouring down my face, working hard over the twisted rock. It is very interesting, very dangerous walking.

My goal is to find a secret oasis deep in the lava flow; a cool green protected place where water pools and flows over the black rock, a place where big trees grow. From what I have heard and read about the area, there are lava tubes, ice caves, and weird volcanic cones to be found out here. There are a few trails, some new and some very ancient, that cross the lava. There are Kipukas, undisturbed areas that the lava flowed around and did not bury, in some cases creating beautiful islands of pine set in the middle of the black lava. Or so I've heard.

Working my way north, I find interesting pockets of grass and trees that grow in protected grottos that are separated by great domes of glassy black rock where nothing lives. No oasis, though: no water. The usual aridity of the Southwest holds true here; this area gets a scant ten inches of rainfall per year. The trees are mostly tough, stunted juniper. Gnarled, twisted, wild-looking things. I love trees, especially the ones that look like they must work for a living, and I often pause to admire a particularly spectacular specimen. Here is a small one, draped in rather athletic fashion across a lava dome, roots running along the surface in sinewy loops before diving straight down a crack. Here is a startlingly large juniper, stubbornly wedged between massive slabs at the very bottom of a sandy hollow. All water that falls into that hollow is hers, it would seem.

A few inches of rain a year and enough windblown, sandy soil to dip a root into, and you have a juniper oasis. There is an unexpected amount of life in here, despite the cruel and bizarre geology. Crouching and crawling, I stop to examine small wonders, marveling at tiny purple flowers, vivid against the lava, that my weak plant taxonomy won't allow me to identify. Lizard tracks make their quick, scrabbly way across a small dune. Deer and coyote prints were plentiful near the transition zone where the lava meets the pinyon-juniper forest, but fifty or a hundred yards into the lava and they began to thin and then disappear. Squirrel tracks circle a tiny, warped pinyon pine. Upright again, I walk slowly, and very carefully. The black rock is formed into fantastic shapes as yet unbent by erosion. There are caves of shiny glass, huge blowholes, and open, broken lava tubes. Walking here is like surfing crested waves of frozen stone.

As I venture further and further from the highway, the lava gets deeper, until, judging from my height above the surrounding terrain, it is hundreds of feet thick. As it cooled, the lava flow fractured and bent. The lava domes and humps are laced with cracks; some wide enough to be a barrier to travel, and my path is often determined in the short term by the direction of the cracks. But north is my direction, and to go north I am forced to cross the crevices. Some of my leaps are small, and others require athletic bounds across cracks fifty, a hundred feet deep. Very thrilling; on these leaps my heart rate goes up to about three hundred beats per minute. And also, terrifying: one slip of the little feet and I am a dead hiker. Even if I were to survive the plunge, climbing the glassy walls would be nearly impossible. A small voice of reason whispers, "no one would ever find you...die down there in the darkness...alone...afraid..."

This is insane. I should turn around, get off the lava and spend the day hiking the slopes under the sandstone cliffs – that would be interesting! I could just go home; my dog misses me...

An eerie, lilting chuckle rolls down in my throat and the little sane voice starts screaming because it knows what's coming. I can almost always tell when it's going to hit. It's the same set of sensations every time. A weird buzzing in my ears, stomach sick and hollow and my body pumped full of sunlight. Everything is clear, vividly outlined, and time seems to creep along; there are no boundaries between myself and anything else. Once, while climbing in the Salt River Canyon, this same extraordinary sensation came over me to the point where I felt like part of the rock, and I went down a terrifying, nasty pitch at a trot. My hiking companion, watching from above, said he thought he would just wait a bit to compose himself before searching for my body, but I arrived at the bottom screaming and exultant, untouched.

Moments ago, I was cautious and deliberate, thinking my way along. Power is in me now; the once-plodding hike turns into a rushing, exhilarating blur. The world at my feet becomes the only one that matters. Totally in sync with my body, mind empty, I flow over the rock like water, gliding, jumping, and trotting, every minute full of a hundred split-second decisions. Blur of crevice, sweat, heartbeat, steep slope, sidehill, jump, slide, trot, reverse direction, build up speed, jump...

§

In the early afternoon, drained by the morning's frolic, I catnap under a lonesome bonsai of a pinyon pine and lunch on sardines and cheese. The tree stands alone near the top of a massive lava dome, giving me a full 360º view, which would be superb if it were not for the heat waves, which twist and blot the scenery in all directions. I do, however, have amazing sights of enormous, shimmering lakes, which shake and shift with the hot breeze like a vision from some wild peyote dream. Lava domes jut from cool waters and junipers stand in a glowing flood. What I see is not water, of course, but the fabled mirages of the desert. A red-tailed hawk, perhaps launching from the high sandstone escarpments to ride the thermals, glides slowly overhead, hunting. Good luck, buddy. Wake me up in an hour or so, and we'll go fishing in those lakes over yonder...

The thin bed of needles shed by the tree makes a welcome cushion. Head propped against the rough trunk, I sleepily survey my remarkable surroundings. Everything has an old, weathered feel to it out here—this tree, the sandstone cliffs, and the jagged wilderness of black stone. From a geological standpoint, however, the lava is young. Volcanic vents opening in the ancient plain, now partly buried, spewed forth a series of lava flows. Some date back hundreds of thousands of years, and some, like the McCarty's flow I am most likely resting on, are estimated to be as recent as 3,000 years old. The surrounding terrain is much, much older, the sandstone mesas a mute display of hundreds of millions of years of sedimentary history. Ancient seas deposited layers of sand and mud, and then receded, the waters advancing again after a time. The earth clenched her fist and young mountain ranges pushed through the older layers of rock, with wind and rain doing their usual patient sculpting work throughout.

Geological time is too vast for my temporary sentience to grasp; the planet seems eternal to me, small life forms like myself flitting about on her belly for an instant before gratefully expiring, exhausted and confused.

After soaking up a few quarts of water, I crimp on my greasy hat and continue the march north. The sardines school and dart madly around my guts as I stagger across the black lava, searching for a cool green oasis, a hidden spring, life on the moon. Heat waves erase the giant bluffs to the south, the way I have come. I consider myself to be a hot-weather hiker, a desert rat, and take a bit of pride in not being one of those wimps that need perfect conditions before venturing out-of-doors. Father Sun is giving me a beating today. All morning, as I gaily skipped and jumped my way across the lava, the jet-black rock had been steadily and efficiently soaking up solar energy. By afternoon the rock is too hot to touch and the desert sun is directly overhead. The effect is intense. Soon, I am walking along with a canteen in my hand, sipping as I go. Under conditions like these, accompanied by heavy exertion, the human body can sweat out nearly a gallon of water an hour (my personal observations). The rattling of empty canteens in my pack reminds me that the two-plus gallons I started with has been mostly consumed, so I just ease along, pausing to lower my body temperature in the shade when I can find it.

I hop across more of the horrible cracks, each deeper and blacker than the last, and finally the sphincter begins to quiver, the heart flutters, and right out there in the middle of the forsaken place I lose my damned nerve. The power of the morning is gone, leaving me with nothing but dread and despair. The sane voice is back, crying in my mind, forcing me to acknowledge (You fool!) the entire humiliating and yet familiar array of physical sensations: throat dry, knees weak, hands clammy. The sardines are making like salmon, leaping up my throat, determined to see the sun once more.

Lying down on a patch of sand (the rock is too hot to touch) at the edge of the Mother Crack, I hang my head over the edge to look straight down into the darkness. Drop a small rock: no sound returns. I surmise from this experiment that there is no bottom to the Mother Crack.

Why do I do these things, when they in fact scare the crap out of me? I am not the type to scale Annapurna or rappel down a five-hundred-foot cliff, or so I say. But I have done hundreds of these lethal little outdoor adventures. Any one of them could have, and this one might, get me killed or at least solidly maimed.

Despite hundreds of warnings along the line of, "That's really dangerous," and, "You shouldn't go out there alone!" I have been solo hiking all my life. Being alone out here is something I enjoy. It's quiet. I can do whatever I damned well please without having to compromise, discuss, consult, or negotiate. No one knows where I am, not a single soul on earth. There is no safety net. If I croak, my bones may not be found for decades. My own wits and physical strength will have to suffice. I like the idea of not being in a position to be rescued; it's my own damned business. This is my pilgrimage—no spiritualist beside me, no teacher to guide me, I just wander, burnt by the sun.

Someone told me a long time ago that you haven't lived until you've cheated death at least once. And those who distain adventure often maintain that extreme adventurers, like climbers, all have a death wish, and their reasons for being on those rock walls have little to do with thrill, or challenge, or breakthrough. They have dark, deep reasons for climbing, and few climbers are aware of these subconscious drives. They climb with Thanatos, the death instinct.

Perhaps. Wilderness is definitely a suitable arena for those who want to flirt with death. What better backdrop for your passing than the face of God? Better by far than beneath the fender of a delivery truck on 43rd Avenue. Personally, I don't think that most daring people are out looking for death. In fact, it is the ancient blend of adventure (extreme sport) and the indisputable power of untouched nature that brings us to life. Other than wilderness, what is the alternative? Slow, numbing death in a noose of information, concrete, organization, pollution, taxation and technology. Sun-blasted, ragged, and wild-eyed, you will find the faithful seekers out here, in the forgotten places. But you must search hard to find them, and when you do, they will scurry off into the rocks before you can speak.

Noble thoughts at least temporarily forgotten, I solemnly consider barfing. Return something to the earth. Besides, the sardines deserve it. Cooked from both sides like a roast in the broiler I lie there for what seems like hours, quietly considering the effects of heatstroke on my balance. Then, without warning, a tiny feather of cool, delicious air wafts up from the subterranean depths of the Mother Crack, caressing my face and down the collar of my shirt. An omen. Knees firm, I come slowly to an upright stance. The hot water from one of my bottles revives me. Sphincter tight, I take two steps back, run, leap, and clear the Mother Crack with a foot to spare.

To hell with it. I am weary of the cracks and have given up all hope of finding the cool oasis of my dreams. There may in fact be such a fantastical place out here among the flows, but without maps or direction a person could spend a lifetime searching. Besides, my boots are rapidly wearing thin on the sharp rock. Hoping to find a trail, I beeline it east off the flow.

It takes an hour of careful maneuver and a few more death-defying leaps to reach the eastern edge of the flow. At the edge of the lava field, a long, slightly convex slope of smooth lava takes me neatly down onto the sandy desert plain. When I pause to look back at the lava flow, it stretches as far as I can see to the north and south like a frozen black tsunami.

After that comes the ritual blessing of the hike: "Damn, that was crazy, but I'm sure glad I did it!" and after that comes the ceremonial plod across miles of gnat-infested burning sand flats. To those who have never experienced a summer hike across the juniper desert, close your eyes and imagine the following, truly sensual experience: You walk under the blazing sun until your head feels like an overripe melon. Sweat stings the eyes, running like a tiny stream from under the foul hat. You take the hat off often to wipe your brow as you slog through the sand. After an hour or so of this, the trusty feet start to do stupid things, like hit each other a lot. You lie down gratefully in the warm sand under a canopy of juniper. Deep, deep shade. It is close to a hundred degrees in the shade, but simply escaping the sun is a minor bit of ecstasy. You work a water bottle out of the pack, and take a long, lukewarm draught. You sink back slowly, gratefully, pulling the hat down over your face...Yesssss.

Right then is when the little bastards hit you. Tiny red-hot coals under your shirt, inside your ear, on the oh-so-tender corners of your eyelids. Hundreds of them, all creating an utterly maddening whine around your ears as they dive upon your supine flesh, each craving the fluids in your body. With an insane snort and bellow you stagger back out into the sun and break into a shuffling lope. The gnats can't bite what they can't catch.

Back at the parking lot, I reward myself for my labors with a short and shady stroll up the nice trail to La Ventana, a gorgeous tiger-striped sandstone arch. Sitting next to the arch in a pool of deep shade with the setting sun highlighting the crumpled skin of the Malpais, it's time to start thinking about finding a place to camp for the night.

And then where will the winds take me? Shiprock? Capitol Reef? The soaring Aquarius Plateau? Or perhaps the weird badlands northwest of the Abajo Mountains... It doesn't matter. As long as I'm out West and out of the city, I'm happy.

Gotta move, the maps are calling me: Kaiparowits Plateau, Fiftymile Mountain, Gunsight Pass, Scorpion Gulch...

BIBLIOGRAPHY

Apter, Michael J. *The Dangerous Edge*. The Free Press, A Division of Macmillan, Inc., New York, 1992.
Freud, Sigmund. *Civilization and its Discontents*. W.W. Norton and Company, New York, 1961.
Robinson, Sherry. *El Malpais, Mt. Taylor, and the Zuni Mountains: A Hiking Guide*. University of New Mexico Press, Albuquerque, 1994.
Schulteis, Rob. *Bone Games*. Breakaway Books, New York, 1984.

Modern Man and the Ancient Ones

One hot summer morning, I find myself in Shiprock, New Mexico, standing in a windswept parking lot and sipping bitter truck stop coffee.

Shiprock. The towering spire for which the town is named rises to the southwest like a colossal sailing ship, her topsails soaring two thousand feet above the desert floor. Several years ago, my wife and I, along with another couple, grilled steaks at her feet.

It was early evening when we arrived. The sky was fantastic. It was the height of the summer storm season, with rain and lightning dancing on the northern horizon, backlighting the terrible majesty of Shiprock. Geologists, I suppose, would label Shiprock a volcanic neck, the hard core of an ancient volcano. A solid mass of stone, the spire leaps out of the flat desert so dramatically, the sight never fails to make my breath catch in my throat. The sheer, harshly cut faces of the tower are so raw, so wild, that out of the corner of my eye I can believe that it moves. Twice as tall as she is thick, the twisted igneous rock leaps into improbably sharp battlements, leaning outward near the summits, dark horns ruling over thousands of feet of nothing. Great sails catch the endless desert winds. She goes nowhere, of course, her stony feet rooted in the very bowels of the planet, and it is the dry wind that must yield.

That day, years ago, the wind was hard and strong and the tremendous hissing roar it made as it swept around and through the immense rock was overwhelming, even intimidating. It sounded as though Shiprock was sailing fast through a foamy sea.

Truck Stop Coffee. (Illustration by Jan York)

§

Rolling north from the town of Shiprock on Highway 666, the high empty desert of northern New Mexico becomes the high empty mesas of southern Colorado. I refuel in Cortez, then drive east toward my destination, Mesa Verde National Park. Now that I am near, I find myself undecided as to whether I want to visit the Park or swing west into Utah. I rarely visit National Parks, approaching them with mixed emotions. On the one hand, National Parks are where some of the nation's most spectacular scenery and wildlife are to be found. On the other hand, they tend to be teeming with people, and crowds are in direct conflict with what I want from an outdoor experience. National Parks: our national enigma.

Dallying on the way, using any excuse to take a roadside rest break, Mesa Verde National Park nonetheless eventually heaves into view. Getting in line with the motor homes and travel trailers, I eat exhaust, pay the fee, smile at the nice lady, and take the packet of map and brochures. The fee, I remind myself, is not a personal commitment to stay at the Park. This is only a reconnaissance.

The large, well-planned campground turns out to be a pleasant surprise. There is plenty of room. It's quiet. I find a nice secluded campsite and begin making lunch in a soft, cool drizzle.

Cleaning up after lunch, I take a few steps into the trees to empty the dregs from my coffee mug and am surprised to see a deer lying not fifteen feet from my truck.

She watches me carefully. Obsidian eyes. Dark, dark eyes, like wild globes of light. The narrow jaw chews steadily, but the subtle tensing of her rear legs tells me I am too close. Fifteen feet was just fine, it seems, but ten feet is too close. The slender legs start to tuck in, and I back away. She relaxes. I relax. I'm OK. She's OK. I lean against the tailgate and study her. The opportunity to watch a wild animal up close is a rare thing. I usually watch deer from across a sizable canyon, as they in turn watch me: look, look at the noisy, tall frail thing. See it sweat. Hear it pant.

The doe lies quietly in the trees, brown and black. Years ago, while hiking in the Sonoran Desert north of Tucson, I took a mid-afternoon nap on the side of a small ridge. The weather was perfect for sleeping, warm and sunny, with a small breeze gently shaking the branches of the Palo Verde trees. I slept hard for a couple of hours, woke, and sleepily climbed up and crossed the ridgeline. Lying just on the other side of the ridge was an adult doe, fast asleep. The magical afternoon had worked its spell on the deer as well. I stood absolutely still for a long time, perhaps five minutes, just watching the sleeping deer. I was so close I could have taken two swift strides and leaped upon her like a mountain lion.

I imagined doing this. The poor startled deer, my arms wrapped around her, holding her down. Deer look rather fragile with their skinny legs and their big ears, but if you have ever seen a deer run, and I mean really run flat out, then you know you have seen strength, for speed is born of power. As I stood watching the doe, I remembered a long line of running deer, out in the acacia flats west of the Dragoon Mountains, flowing over a barbed wire fence in the fading purple light like water over smooth stone.

Fantasy aside, I then imagined the deer leaping up and pounding the hell out of me, the iron muscle exploding out of my grasp, me upside down in the prickly pear hoping she won't come back for more. After a while I just crept away, feeling clever for having counted coup on the wily desert deer.

Wild animals, no matter how calm and tame they may appear, can be exceedingly dangerous. Not anxious to press a frightened deer, I cross the road to empty my cup, thinking about the camera resting on the dashboard of the truck. No. Photos of wild animals that have become accustomed to the proximity of humans look about as exciting as photos of zoo animals. There is no thrill of the hunt.

I spend the next hour playing solitary tag with tourists in the big National Park campground. Here is how you play solitary tag: find a secluded campsite well away from everyone. Set up all your stuff (this is important). Wait. Without fail, a towering motor toilet will pull up in the space right next to yours, ignoring ten empty spaces just down the road. It happens with unerring regularity, no matter where I travel, and I still wonder why. Perhaps it is because my camps appear to be solid, secure refuges in the wilderness, exuding an aura of safety. People must look at my camp and think, "Now here camps a solid citizen. And a rugged outdoorsman, as well."

I was once camped deep in the forest south of the Grand Canyon. It was high summer, and all the roads cut through the forest were dry and passable to nearly any vehicle. Hundreds of square miles of empty country, a great deal of which is prime camping area, surrounded my isolated and unobtrusive little camp. After making camp (which consisted of setting up my small tent and tossing the ice chest under a tree), I went on a short hike into the woods to look for antelope. Returning a couple of hours later, I found my camp absolutely surrounded by chattering tourists from Ohio. Not just surrounded, but occupied. One couple had pitched their tent fifteen feet from mine!

Perhaps it is simply herd instinct. Safety in numbers. Bears in the woods. Pain in the ass. Move camp. Wait ten minutes...

Over the next few days I move my camp four times.

With rain threatening and thunderous clouds swirling around the mountains, I grab my trusty daypack, strap on my boots, and spend the day hiking up, across, down and around a beautiful mesa that overlooks the campground. The day is warm and humid, hot in the sun and cool in the shade. Despite the horde below, this mesa belongs to me. Want to leave the crowds? Take a deep breath and step off the pavement. Want to avoid people altogether? Get off the trail and plunge into the brush. You may get scratched and bloodied a bit, and you might get lost (No engineered walking surfaces, road signs or helpful rangers.). And there is the possibility of death; the broken leg, the horribly swollen ankle as the rattlesnake watches you from under a rock, the ominous rattle of empty canteens as your blackened tongue fills your mouth – but survive and take home an uncommon awareness instead of a damned pamphlet.

The mesa is artistically cut with a series of small box canyons, each flat-bottomed and neatly capped with hard white sandstone. At the head of each of the canyons, I pause to sit and gulp in the scenery like a starving man. Which indeed I am; I have been too long in the city. Too long trapped by walls and concrete. Cities slowly starve me, bringing on a gradual, so slow as to be nearly unnoticeable, unease and discontent. The soul churns, twisting, twisting. It is a terrible, wretched condition, which can only be cured with distance, space, time, rock, and silence. It makes me wonder if I have made the proper choices in life.

Born free, I was. And then forced to work. And cities are where the manufacturing facilities are. Factories which produce a variety of indispensable gadgets, requiring engineers with my training and skills to design and build. I could live in some isolated setting and have all this at my back door, but live a more frugal life and probably engage in an occupation that would quickly bore me. I have exchanged that which I love, the wild places, for other things.

The peace of the summer afternoon is briefly torn by the mad charge of a thunderstorm. I have left the faint deer trail. It doesn't lead where I want it to, so I go my own way. I want to sit on the very end of a crag that I noticed earlier in the day, a point of bare rock that hangs over the confluence of two canyons. Through the binoculars, there is a spot on the very tip of the point that looks ideal for solitary contemplation. One lone pine, twisted and tortured by wind, snow, and rain, stands on a barren apron of white rock. Its stance speaks of suffering, of silence, of unquenchable life.

The tree commands an enviable view. Blue mesas, tilted and tabled and rolling and running for twenty miles, and beyond, a dark hint of river valley framed with a golden, slanting curtain of rain, and further still, not clearly seen through the mist but surely there, the massive wall that is the Rocky Mountains.

I have often been jealous of such trees, growing in a place that is the center of light and distance, queens of sweeping panorama that I can enjoy only briefly. Year after year, decade upon decade, perhaps existing for living centuries, standing sentinel over such beauty. Trees like that live in my mind forever. No matter how frenetic and foolish my animal life becomes, vibrating comically, chaotically toward death, the clear memory of a tree on a craggy ridge, or a great saguaro atop a desert hill, brings me peace.

Taking to the trees, I pick my way along sandstone outcroppings free of the clinging brush. My destination is west, but the route is never a straight line: north for fifty yards to clear a field of boulders, then nearly west for a hundred yards or so, a detour south to miss a stand of thick brush, a quick slide down the slope of a west-sloping ravine, back on course. Through a gap in the trees, I get my first look at the storm. It is coming, but it looks like it will pass well to the west of the mesa.

I continue on, and the top of the storm cloud, bent over by high altitude winds, sweeps overhead and crosses the sun. The wind, which has shifted randomly all day, now comes steadily from the west as I press through the cedar.

The gnats take a break. This is excellent, as the tiny fiends have been rattling around in that space between eyeballs and sunglasses like miniature handballs. The wind turns cool and starts to howl. Fantastic. The storm is going to pass the mesa closer than I had guessed, and it is coming fast. I start to jog along, weaving through the trees, wanting to be there in that perfect place when the storm passes. Thunder rolls, then again, and I break into a run, the heavy binoculars banging against my chest and the pack riding too low.

It is further than I had thought. It always is. Ten thousand miles of hiking, and two miles still looks like one and a half. Lights dance in front of my eyes. Blood sugar low, and dehydration is making me miserable. No problem, I'm almost there, and then I will indulge in lunch and a thunderstorm. The breeze is delicious.

I am tired and ragged by the time I reach the end of the long arm of rock. The place is perfect; a bed of soft needles awaits me under the tree...

Suddenly, lightning strikes on the south end of the mesa. It is, I realize, not a large mesa. The storm track has changed and here it comes, and this is not a good place to be.

Running up and down along the cliff, I look for a break in the caprock, a way off this lightning target. If there is, I will climb down and let the storm roll over me. No luck. The caprock stands firm, not a break in sight along the hundred-foot cliff. Dark cloud and rain move rapidly toward me down the long line of cliff and canyon.

Time to go. I beat it the hell out of there, back across the bare stone, low blood sugar replaced by gallons of adrenaline. "Gobabygobabygo! Gobabygobabygo!" I chant merrily as I run. Diving into a likely stand of low-growing cedar, I let the storm pass, cringing as the lightning bolts work across the mesa.

Ten minutes later, the sun is out, the storm is moving east, and it is steamy hot.

The evening is a beautiful one, cool, with lightning flashing on the distant peaks of the Rocky Mountains. I amble around camp, enjoying the slow, deep bliss that accompanies not having a damned thing to do or anyplace in particular to be. I prepare a simple meal of steak, bread, green peppers and jalapenos, and a cold brew. As I eat, enjoying the shadows on the mesas, deer and turkeys come right into my camp, just passing through.

The next day, I pack up and do the Park, the Visitor Center, the whole tourist package. At the Visitor Center I get in a long line that begins outside the building. We patiently file in, where we learn two amazing facts:

1. Within the park there are over four thousand sites that archeologists have identified as being Native American.
2. Within the park there are what appear to be over four million mildly irritated tourists searching for a restroom.

According to figures the Park publishes, a sizeable portion of the planet's population visits the Park on a daily basis. Of course, that may not be entirely accurate, as some of the people are regulars, with yearly passes and that sort of thing. There may in fact be primitive peoples living in a rain forest somewhere south of the equator who have never visited Mesa Verde National Park. These people probably wouldn't be interested in the Park, as they themselves are most likely tired of being visited by archeologists, having their photos taken, and having their ancestors exhumed for examination.

Modern Man continues through the Visitor Center, collecting maps, brochures and other general park information. We exit the building and return to our cars. Then we all get in line on the road to view the ruins.

The ruins are superb. I have seen photos of the major sites thousands of times, but to see them in their natural setting is a moving experience. The most common construction material of the Anasazi builders was sandstone, shaped into rectangular blocks. The dwellings blend with the sandstone cliffs and caves so well that at a distance it is difficult to tell where the cliff ends and a human-built wall begins.

My frustration with the National Park Service fades as the day passes. Of course, these important dwellings must be carefully protected, or they would soon be destroyed. It is not the fault of the Park Service that the crush of visitors grows yearly. I blame the Motorized Tourist (of which I am one) for the crowds.

Near the end of the day, I walk down a gently graded paved trail to view a site that is, like many in the area, built into a shallow cave. A few minutes earlier, there were dozens of people down here, but they have all returned to their cars. There is no one in this narrow canyon but me. Standing in the shadowed silence, I try to imagine life here, what it would be like to live here, work here, die here. It is very difficult. Hand-and-toe-holds carved into the rock cliff connect the cave to the canyon floor. Only when I imagine children's voices calling down from the high rocks does my fantasy spring to life.

When I return to the top of the canyon, I leave the silence and the Anasazi behind. The mesa tops echo with the roar of automobiles. Soft, weak bodies that require a paved trail have replaced lean, hard bodies that could go up a cliff like a spider.

Even though the Park is dedicated to the memory and deeds of the Anasazi, the Ancient Ones are gone. Tough, strong and quick, they made a living from this hard land. Why they left is still a mystery. The busy communities and the farmlands are all empty, leaving a dusty sadness that hangs in the clear air like a shadow seen from the corner of the eye.

I'll be back, again and again. I can't help it; I have to do this. Ridge, peak and valley, I want to see it all. I want to walk all the wild canyons that smell of hot stone and cool water, where life ancient and new is fed by a living river, and where the wind bears the power of the earth upon its invisible wings.

My children have walked the land with me. If I have done one powerful thing for them, it has been to pass on the gift my father gave to me, which is to reveal the beauty that exists in all things. It has not been a difficult thing to do. Children are full of magic. As long as I can walk the wild lands, I will, and when I can no longer walk, I'll make my children take me out to the wild places so I can sit and listen.

Petroglyph. (Illustration by Jan York)

Other Books of Interest from Sunstone Press

Barker, Elliott S. *Smokey Bear and the Great Wilderness*. Santa Fe: Sunstone Press, 1982.

McDonald, Corry. *The Dilemma of Wilderness*. Santa Fe: Sunstone Press, 1986.

_____. *Wilderness: A Guide to Wilderness Areas in New Mexico*. Santa Fe: Sunstone Press, 1985.

Nusbaum, Rosemary. *Tierra Dulce: Reminiscences from the Jesse Nusbaum Papers*. Santa Fe: Sunstone Press, 1980.

Prisciantelli, Tom. *Hiking North America's Great Western Volcanoes*. Santa Fe: Sunstone Press, 2004.

_____, *Spirit of the American Southwest*. Santa Fe: Sunstone Press, 2002.

READERS GUIDE

1. Why does the author insist that a hiking guide isn't necessary when heading out for an outdoor adventure?

2. Why is love of wilderness important?

3. What do you think of the author's claimed addiction for wilderness? Do you think it is it real, or just an affectation?

4. Why is public lands cattle ranching a such a passionate subject? Name two ways that cattle ranching on public lands can harm the local environment.

5. Why is wild game hunting often a focal point for passionate debate?

6. In "Rite of Passage," why is the "line in the sand," referring to the killing of living things for food, so variable from person to person, and from culture to culture?

If you are vegan or vegetarian, skip questions 7 and 8:

7. Why does the perceived value of life play an important part in what we choose to eat? Name two animals/fish/fowl you eat regularly, and two you would never eat due to their high value.

8. Does the supply (current abundance) of an animal/fish/fowl play a part in your decisions regarding what you eat?

9. Do you have a fear of bears, even if you are unlikely to ever encounter one?

10. How does poor planning play into outdoor adventure? Or any adventure (such as travelling by air to a place you have never been)? Do you think most "adventure" can be avoided by meticulous planning?

11. Do wild rivers still have a place in our increasing modified landscape? Name two benefits of a wild river as opposed to a reservoir.

12. In "Lava and Sunlight" there is mention of Thanatos, the death instinct, and the ancient blend of extreme sport and the indisputable power of untouched nature. Do you think these activities are a "death wish," or an affirmation of life?

13. When National Parks are so popular, why does the author claim they are in conflict with what he calls "uncommon awareness?" Name two ways that National Parks could be improved.

14. How has industrialized tourism impacted National Parks?

www.ingramcontent.com/pod-product-compliance
Lightning Source LLC
Chambersburg PA
CBHW011802190326
41518CB00018B/2570